# Rebalancing Our Climate

# Rebalancing Our Climate

## The Future Starts Today

Eelco J. Rohling

# OXFORD
## UNIVERSITY PRESS

Oxford University Press is a department of the University of Oxford. It furthers
the University's objective of excellence in research, scholarship, and education
by publishing worldwide. Oxford is a registered trade mark of Oxford University
Press in the UK and certain other countries.

Published in the United States of America by Oxford University Press
198 Madison Avenue, New York, NY 10016, United States of America.

Library of Congress Cataloging-in-Publication Data
Names: Rohling, Eelco J., author.
Title: Rebalancing our climate : the future starts today / Eelco J. Rohling.
Description: New York, NY, United States of America : Oxford University Press, [2022] |
Includes bibliographical references and index.
Identifiers: LCCN 2021009734 | ISBN 9780197502556 (hb) |
ISBN 9780197502570 (epub) | ISBN 9780197502587
Subjects: LCSH: Climate change mitigation.
Classification: LCC TD171.75 .R64 2022 | DDC 363.738/746—dc23
LC record available at https://lccn.loc.gov/2021009734

DOI: 10.1093/oso/9780197502556.001.0001

1 3 5 7 9 8 6 4 2

Printed by Sheridan Books, Inc., United States of America

*To the next generations—*
*that they may prosper sustainably*

# Contents

# Acknowledgments

This book was written intermittently between August 2019 and October 2020. The entire period overlapped with convalescence following an ankle-repair operation, followed by Australia's terrifying bushfire season, and then the coronavirus pandemic. It was an eventful time, with one clear common denominator: lots of working from home. I thank my family for support and for tolerating my reduced participation in regular family life, and my parents and brother for supporting all the choices I have made, no matter how irrational they may have seemed at the time. I gratefully acknowledge support and advice from my fellow "climate-solutions enthusiasts," Henry Adams, David Beerling, Justin Borevitz, Aurore Chow, Kerryn Brent, Phil Boyd, Wolfram Buss, Pep Canadell, Rebecca Colvin, Steve Eggins, Michael Ellwood, Jim Hansen, Will Howard, Andy Lenton, Adrienne Nicotra, Jan McDonald, Jeff McGee, Cameron O'Neill, and Kirsty Yeates. I am grateful to countless colleagues and friends for discussions and collaborations over the years that helped shape my views across a wide range of ocean and climate research, notably: Ayako Abe-Uchi, Sheldon Bacon, Edouard Bard, Andre Berger, Henk and Dan Brinkhuis, Wally Broecker, Harry Bryden, Ken Caldeira, Tom Chalk, Peter Clark, Rob DeConto, Michel Crucifix, Jerry Dickens, Harry Elderfield, Fabio Florindo, Gavin Foster, Simone Galeotti, Tom Gernon, Phil Goodwin, Katharine Grant, Ivan Haigh, Julia Hargreaves, Gabi Hegerl, Jorijntje Henderiks, Dave Heslop, Anna von der Heydt, Fiona Hibbert, Matt Huber, Peter Huybers, Peter Köhler, Gert de Lange, Juan Cruz Larrasoaña, Gordon Lister, Lucas Lourens, Dan Lunt, Gianluca Marino, Valerie Masson-Delmotte, Paul Mayewski, Taryn Noble, Joe Ortiz, Bette Otto-Bliesner, Heiko Pälike, Gert-Jan Reichart, Andrew Roberts, Laura Rodríguez-Sanz, Gavin Schmidt, Nick Shackleton, John Shepherd, Steven Sherwood, Mark Siddall, Appy Sluijs, Axel Timmermann, Paul Valdes, Bert Vermeersen, Roderik van de Wal, Felicity Williams, Jimin Yu, Jan-Willem Zachariasse, Jim Zachos, and Richard Zeebe. I thank Caroline Dycke for extensive help with conveying the message without drowning it in technical detail; Larisa V. Medenis for much-needed help with illustrations; Henry

Adams, Tom Goreau, and John Shepherd for providing super-constructive reviews of the first complete draft; David Orr and an anonymous reviewer for support and suggestions to the initial book proposal as well as the final manuscript; and Jeremy Lewis for managing the project at Oxford University Press.

# Prologue

For much of the time I was writing this book in Canberra, Australia, during our summer of 2019–2020, bushfires raged throughout the country. Great swaths of forest were burned, including temperate rainforests. Smoke caused cities in Australia to rank as the most polluted in the world, by some margin. In the end, more than 12 million hectares had been burned, most of which was native forest and woodland that supported unique ecosystems, along with vast tracts of private land, including agricultural assets.

More than 1 billion or more mammals, birds, and reptiles were lost, along with innumerable bats, frogs, and insects and other invertebrates (Elsworthy, 2020; Lewis, 2020). The fires ravaged villages and towns, roads, power substations, communications infrastructure, bridges, cultural heritage sites, and so on. Some 2000 houses were destroyed, there were more than 30 casualties, and thousands of people were evacuated and displaced. Emergency services—most staffed by volunteers—had been worked to exhaustion. By early January 2020, the costs of recovery from this crisis were already estimated at $70 billion (Quiggin, 2020), and new fires were still appearing through the end of February.

At the end of January 2020, a major new fire developed that rapidly closed in on Canberra, the nation's capital and home to about 400,000 people (Figure P.1). The most immediate threat was to properties on the city's outskirts, like mine, which backs onto a nature reserve with nothing other than woodland between us and the fire front. But the Canberra fires of 2003 demonstrated that fire can easily penetrate deep into the city's heart. Conditions this time were perilously similar. Fortunately, history was not repeated because a change in the winds, and some heavy rains a week later, allowed the fire services to gain the upper hand.

Why did this unprecedented fire season happen? Fundamentally, the 2019–2020 bushfire crisis was a "weather-driven event, underpinned by years of warming and drying [because of climate change]" (Mullins, 2020). Ambient temperatures had been breaking record upon record, amid a succession of ever more numerous, intense, and persistent heatwaves

**Figure P.1** The Orroral Fire in Namadgi National Park to the southwest of Canberra, which came within about 13 km of the author's house (January 28, 2020). Eventually, a shift in the wind direction, followed a week later by heavy rains, allowed the fire services to slowly gain control and extinguish the blaze through February 2020.
*Source*: Photo by the author from just outside his back garden.

(King, 2017), and there had been severely reduced rainfall for two or three years. Everything was dry as a bone, and dry undergrowth and leaf litter had accumulated rather than decomposed; every single spark could ignite an uncontrollable blaze. Thinning out of forests and controlled hazard-reduction burns in open forests proved to be of limited preventive value, given that the fires moved through the forest canopy and easily skipped through treated areas (Hayman, 2020). In fact, opening forests may promote the drying out of forests and soils, making them more susceptible to fire (Mottram, 2020).

While individual fires cannot be attributed to climate change because there is a range of underlying causes (Nguyen et al., 2020), the unprecedented intensity of the 2019–2020 bushfire season had been long predicted, based on a clear relationship between climate change and the probability of long, intense fire seasons. In 1988 and 1995, the meteorologist Tom Beer had revealed a trend by which "temperature goes up, it gets drier, and then the fires go up" (Readfearn, 2019). And in reviewing the impacts of climate change on Australia and its economy, the 2008 Garnaut Climate Change Review highlighted "that fire seasons will start earlier, end slightly later, and generally be more intense," and specifically predicted a more frequent and intense fire season by 2020 (Baker, 2020). Unfortunately, similar arguments made the media frequently another time in 2020, with the September bushfire catastrophe in the western United States.

The bushfire crisis made headline news around the world for several months and intensified the international public cry for action to curb climate change. Unfortunately, it may be impossible on short timescales to stop ongoing and looming fire crises in Australia and elsewhere, including the west coast of the United States, Siberia, Mediterranean countries, and countless other places. But on timescales of decades, something can certainly be done about the underlying drivers:

> Global warming is related to the total amount of greenhouse gases accumulated in the atmosphere. Even without any new emissions, the legacy of already accumulated global emissions would maintain temperature where it is now. Worse, some additional year-on-year warming would continue because of slow responses in the climate system. To reverse the warming trend, a large net drawdown of carbon dioxide from the atmosphere is needed. Bringing this to the scale required is **humanity's grand challenge for the rest of this century**. (Rohling, 2019)

In short, the timeliness of this book was well illustrated by Australia's apocalyptic fire season, followed within less than a year by similar devastation in the western United States. And across the two was laid the 2019–2020 novel coronavirus crisis, with a global economic impact that was closely associated with a 5–10% reduction in greenhouse gas emissions. The coronavirus crisis acutely highlighted some of the major challenges we'll face in reducing emissions to zero and even into the territory of net greenhouse gas removal.

This book is not about immediate measures to reduce fire risks, or health crises. Instead, it addresses the major underlying trend of climate change and humanity's grand challenge to reverse it. What can we do? And do we have options that won't destroy the economy?

## Prologue Sources

Baker, N. In 2008, the Garnaut Climate Change Review said Australia would face a more dangerous fire season by 2020. *SBS News*, January 7, 2020. https://www.sbs.com.au/news/how-a-climate-change-study-from-12-years-ago-warned-of-this-horror-bushfire-season

BBC News. Australia fires: a visual guide to the bushfire crisis. *BBC News, Australia*, January 31, 2020. https://www.bbc.com/news/world-australia-50951043

Bowler, J. We don't want to alarm anyone, but a large amount of Siberia is on fire right now. *Sciencealert*, May 4, 2020. https://www.sciencealert.com/so-i-don-t-want-to-alarm-anyone-but-a-huge-amount-of-siberia-is-on-fire

Chang, C. How the 2019 Australian bushfire season compares to other fire disasters. *News.com.au*, January 8, 2020. https://www.news.com.au/technology/environment/how-the-2019-australian-bushfire-season-compares-to-other-fire-disasters/news-story/7924ce9c58b5d2f435d0ed73ffe34174

Elsworthy, E. NSW bushfires lead to deaths of over a billion animals and "hundreds of billions" of insects, experts say. *ABC News*, January 9, 2020. https://www.abc.net.au/news/2020-01-09/nsw-bushfires-kill-over-a-billion-animals-experts-say/11854836

Hayman, R. RFS Commissioner says hazard-reduction burns made his organisation "public enemy number one." *ABC News*, January 8, 2020. https://www.abc.net.au/news/2020-01-08/nsw-fires-rfs-commissioner-weights-in-on-hazard-reduction-debate/11850862

King, A. Are heatwaves "worsening" and have "hot days" doubled in Australia in the last 50 years? *The Conversation*, June 19, 2017. https://theconversation.com/are-heatwaves-worsening-and-have-hot-days-doubled-in-australia-in-the-last-50-years-79337

Lewis, S. Over 1 billion animals feared dead in Australian wildfires. *CBS News*, January 7, 2020. https://www.cbsnews.com/news/australia-fires-over-1-billion-animals-feared-dead/

Mottram, L. Hazard reduction burning is not a panacea to bushfire risk: expert. *ABC News*, January 8, 2020. https://www.abc.net.au/radio/programs/pm/thinned-forests-can-be-more-prone-to-fire,-expert-says/11853280

Mullins, G. I tried to warn Scott Morrison about the bushfire disaster. Adapting to climate change isn't enough. *The Guardian*, January 20, 2020. https://www. theguardian.com/commentisfree/2020/jan/20/i-tried-to-warn-scott-morrison-about-the-bushfire-disaster-adapting-to-climate-change-isnt-enough

Nguyen, K. et al. The truth about Australia's fires—arsonists aren't responsible for many this season. *ABC News*, January 11, 2020. https://www.abc.net.au/news/2020-01-11/australias-fires-reveal-arson-not-a-major-cause/11855022

Quiggin, J. Australia is promising $2 billion for the fires. I estimate recovery will cost $100 billion. *CNN Business*, January 10, 2020. https://edition.cnn.com/2020/01/10/perspectives/australia-fires-cost/index.html

Readfearn, G. "What could I have done?" The scientist who predicted the bushfire emergency four decades ago. *The Guardian*, November 17, 2019. https://www. theguardian.com/australia-news/2019/nov/17/what-could-i-have-done-the-scientist-who-predicted-the-bushfire-emergency-four-decades-ago

Rohling, E. Slow the flow and let the soil drink its fill. *The Australian*, December 31, 2019. https://www.theaustralian.com.au/commentary/slow-the-flow-and-let-the-soil-drink-its-fill/news-story/45dbff7b201671051327bb329ff4f025

Thomas, T. Arctic wildfires emit 35% more $CO_2$ so far in 2020 than for whole of 2019. *The Guardian*, August 31, 2020. https://www.theguardian.com/world/2020/aug/31/arctic-wildfires-emit-35-more-co2-so-far-in-2020-than-for-whole-of-2019

# 1
# Introduction
## Outline of the Challenge

Earth is poorly, we humans are the cause of it, and the situation is becoming critical. Humanity's multi-faceted impacts on Earth and its ecosystems pose an imminent threat of extinction to 1 million animal and plant species. This is because of changes in land and sea use, direct (over-)exploitation, climate change, pollution, and invasive species (Intergovernmental Science-Policy Platform on Biodiversity and Ecosystem Services; IPBES, 2019). This echoes warnings from biologists that we're racing into Earth's sixth mass extinction, with modern species extinction rates more than 1000 times higher than natural rates. These rates exceed those of Earth's most dramatic past mass extinction event. The IPBES report also emphasizes that the impacts of climate change range all the way from genetic levels to ecosystem levels and are expected to overtake other extinction drivers during the next few decades. Further key assessments of humanity's impacts, including that of climate change, are given in the Intergovernmental Panel on Climate Change (IPCC) assessment reports, and other United Nations (UN) reports such as the Global Environmental Outlook.

Many of the facets of human impact are closely interrelated and are also coupled to the extraordinarily rapid increase in human population numbers. From less than 1 billion in 1800 and just over 1.6 billion in 1900, the human population has accelerated to just over 7.7 billion in 2019. By today, three-quarters of the land-based environment and two-thirds of the marine environment have been significantly altered by human actions, while crop and livestock production consume some 75% of freshwater resources. Plastic pollution has increased tenfold since 1980. An astonishing 300–400 million tons of heavy metals, solvents, toxic sludge, and other industrial waste are being dumped each year into the world's waters. And fertilizers making their way into coastal ecosystems are responsible for 400 oxygen-depleted ocean dead zones.

We can't claim to be surprised by these findings. Over the last two decades in particular, many scientific, media, and assessment reports have documented the increasingly detrimental human impacts on virtually all aspects of the planet's life and environmental health. Most notably, we are now facing critical consequences from rapidly increasing heat stress and heat death occurrences; sea-level rise; extreme weather events; heat-related ecosystem collapses and critical species migrations; and dangerous feedbacks related to processes such as permafrost decomposition and associated $CO_2$ and methane releases (e.g., McGrath, 2020; Casella, 2020). Finally, let's not forget suggestions that many of the conflicts and refugee streams of recent decades arose at least partly from changing climate conditions and ecosystems, which can drive mass water and food shortages that help trigger unrest and revolts (e.g., Kelley et al., 2015; Vidal, 2019) although the relationship is not settled (Meyer, 2018).

It almost seems as if there is no way out. But here's the critical bit: **Don't give up, because the doomsday scenario really isn't the only way**. Decisive action can still make a significant difference. We can draw inspiration from previous successes. Some came about because of regulation from the top down, as was the case with the Montreal Protocol that banned CFC gases because they eroded the stratospheric ozone layer, which protects life from harmful ultra-violet radiation. Top-down intervention also drove limitations on the use of coal in cities to fight rampant air pollution, which was impinging on public health, and banned lead additions to fuel and paint so as to limit environmental concentrations of that toxic element. In other cases, action gains momentum out of grass-roots public concern and resistance. The ban on whaling is an example of global public outcry driving decisive international action. In the push for action against climate change, a substantial grass-roots movement is currently developing to ensure that international agreements are followed by decisive action. Recent years have seen powerful school strikes and marches for climate action on a global scale, with exceptional speeches by next-generation citizens such as Greta Thunberg, as well as a concerted global call to declare a climate emergency. Thus, the public and especially the younger generation are pressuring the system. Many nations or states/regions within reluctant nations are heeding the call and have declared a climate emergency along with ambitious targets for reducing climate change. It's a great start, but it is absolutely vital that the effort continues with similar energy.

In one book, I cannot tackle the entire gamut of human impacts. Instead, this book focuses specifically on human-caused, or anthropogenic, change in Earth's climate system, including the oceans. I work on climate and ocean change every day, researching natural variability in pre-historic times, before humanity became an important factor. In technical terms, my scientific disciplines of past climate and past ocean research are known as paleo-climatology and paleoceanography, respectively. This background gives me a deep appreciation of the causes, scales, rapidity, and consequences of natural variations in Earth's climate and ocean system, including the life contained therein, and also past mass extinctions. It informs me of how modern humanity's impacts compare with the natural context. I outlined the deep-time natural-history perspective to climate change, from the formation of Earth to the present, in my first book, *The oceans: a deep history*. Humanity's impacts were compared and contrasted with the natural cycles of climate change in my second book, *The climate question: natural cycles, human impact, future outlook*. I am left with no doubt that human actions are at the very heart of today's profound and exceptionally rapid climate change and its impacts, and that the situation has become entirely unsustainable.

When arguing a case for change, however, we have to face the reality that fossil-fuel energy, petrochemical-based manufacturing, industrialized agriculture, and widespread use of cement and concrete have brought massive improvements in living standards, life expectancies, comfort of life, economic outlooks, and so on. This, along with deeply rooted capitalist incentives and strong emotional comfort-zone reasons, creates a major innate resistance to change. Yet, it is equally obvious that we cannot continue as today in the face of relentless human population growth and an associated increase in per capita energy and resource consumption as industrial-economic development spreads. Thus, we find ourselves at a crossroads where we must take the route of sustainable behavior if we, and our children and grandchildren, wish to keep enjoying our time on this planet. After all, we have but one planet, and its ecosystem services are essential to our survival. The business-as-usual approach of stressing the planet's life support systems—its climate and ecosystem services—to the point of collapse is pretty stupid, for lack of a better word.

Allowing the current extinction rates to continue unchecked will severely impoverish Earth's ecosystems before the end of this century. That will decimate food production for human consumption, which we can ill afford because this food production must instead be doubled by 2050. In sum, we desperately need to achieve a more sustainable approach if we are to improve the chances that Earth can continue to look after us. If we don't, then we will deplete all Earth's essential resources and ecosystem services and end up in a crisis during which the planet spits us out. Or, to be more precise, it spits out our children and grandchildren.

And there we have it: we have no choice. Business as usual means betraying our children's and grandchildren's trust that we will leave them a hospitable Earth. It is therefore time for a fundamental change in how we treat our planet, its climate and oceans, and its ecosystems, starting not tomorrow or in 20 years' time, but right now. Every day we dither exacerbates the issues we have to repair.

But what's to be done? What are the options? Many people don't have this information to hand and don't feel comfortable wading through the scientific and engineering literature to build up the understanding. This is why I decided to write this book, to document the surprising wealth of current thought about how we can make a difference.

\* \* \*

Like most aspects of human impact, anthropogenic climate change has to date been clearly related to population growth. In the specific case of climate change, this effect has been multiplied by the growing population's energy-demanding industrial-economic development. It is important to emphasize, however, that these relationships are not obligate; going forward, we can break free of their confines. Along with most other climate scientists, I argue that we should do exactly that and do it urgently. In parallel, we should urgently address and remedy the influences that have accumulated already, over the past two centuries of industrialization and in particular over the six most recent decades.

Scientists and engineers have steadily developed a wide range of tools and concepts to address this challenge, and new ones crop up every day. Yet, the vast majority of these tools and concepts remains stuck in the computer-modeling or laboratory-experiment stages, or occasionally in the small-scale demonstration stage. Progressing to demonstration sites and large-scale test sites is financially demanding, and attracting industrial partners is often problematic because economic profitability is not often evident for pioneering efforts. Regardless, there are significant success stories too, and these often show that increasing implementation drives cost reductions in parts and manufacturing and in many cases also rapid increases in net profitability; for example, consider solar power, wind power, and increasingly also battery storage and transport electrification. Unfortunately, compared with the gargantuan scale of the challenge of reversing the anthropogenic climate impact, these success stories largely remain a drop in the ocean. Much more action is needed.

Reaching an unprecedented level of international consensus about the need for more action, the 2015 Paris Climate Agreement formulates a need to limit warming to within 2°C, and by preference within 1.5°C. It was followed by a wave of national climate pledges for action toward this goal. Despite the pledges, some politicians in small industrialized countries have muttered (especially around election time) that their nations' emissions are tiny relative to those of the likes of China, the United States, and India, so that emissions reductions in their small nations are futile and too expensive. Of course, this overlooks that the large emitters also have larger populations, so that their per capita emissions may be closer to, or even lower than, those of the small nations. More critically, however, such political point-scoring efforts tend to deflect attention from the real elephant in the room, which is the fact that the challenge ahead is so enormous that only a robust joined-up global strategy can be successful.

So, let's view the matter in a more solution-oriented way. As mentioned, the Paris Agreement sets a target maximum of 2°C warming, and a preferred limit of 1.5°C. Incidentally, many climate researchers, myself included, remain uncomfortable with anything higher than 1°C warming by 2100 because slow responses in the climate system will extend warming from that whatever we do.[1] Regardless, the combined national pledges following the Paris Agreement suffice only for limiting warming to roughly 3°C. Even worse, most nations are falling considerably short of meeting their pledges, so that even greater warming might become locked in. Given that most pledges are greatly focused on emissions reduction, it's evident that something more drastic and wide-ranging is needed.

The international research community is converging on what such a drastic and wide-ranging approach might look like. It is a multi-pronged strategy (Figure 1.1). One prong in the strategy for dealing with anthropogenic climate change involves rapid, massive reduction of greenhouse gas emissions (Chapter 3). The other prong concerns implementation of ways

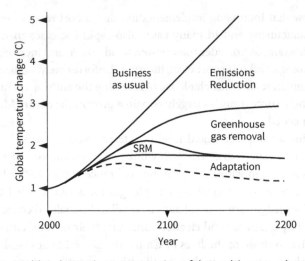

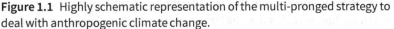

**Figure 1.1** Highly schematic representation of the multi-pronged strategy to deal with anthropogenic climate change.

*Source*: Modified from Long, J., and Shepherd, J.G. The strategic value of geoengineering research. In: *Global Environmental Change, Handbook of Global Environmental Pollution, 1,* Springer, Dordrecht, Netherlands, pp. 757–770, 2014.

to remove greenhouse gases from the atmosphere (Chapter 4). And a further potential prong may be increasing the reflectivity of Earth to incoming sunlight, to cool certain places down more rapidly (Chapter 5). Finally, we need to protect ourselves from climate-change impacts that are coming regardless: that is, the impacts that have become inevitable already (Chapter 6).

Make no mistake: this type of multi-pronged strategy for dealing with anthropogenic climate change is a truly major undertaking. It represents humanity's biggest challenge yet, but it also creates a wealth of opportunities. It is therefore vital that a deep public awareness and acceptance is built of the tools and concepts available for each of the strategic prongs. Without such awareness and acceptance, it will be all but impossible to drive political action toward urgently commissioning the required research and development, including effective concept demonstration and then full-scale implementation. To help in creating this awareness, this book first outlines briefly what anthropogenic climate change is and why it is taking place, in straightforward terms. It then presents current thoughts about how we might minimize anthropogenic climate change, partially remedy it, and address impacts that have become inevitable already.

My two previous books detailed how the Earth System operates, how natural climate cycles developed, how humans are moving the goalposts, and what future consequences may be expected. This book first summarizes key fundamentals about the climate system and causes for its change and then presents an overview of approaches, along with their critical opportunities, challenges, and limitations.

For clarity, I have kept call-outs to specific references within the text to the minimum needed to substantiate specific statements of the greatest relevance. When desired, further source information may be found in the books listed at the end of this chapter and the reading lists provided at the end of each chapter. The reading lists focus mostly (but not entirely) on open-access sources.

The story splits into a series of main components. Chapter 2 provides background on the greenhouse effect and anthropogenic climate crisis, as well as the changing public perception. Chapter 3 considers the urgent need for reducing greenhouse gas emissions. Chapter 4 deals with the additional need for greenhouse gas removal from the atmosphere. Chapter 5 assesses an option that many find controversial, solar radiation management, which does not alleviate high $CO_2$ levels or ocean acidification but may offer rapid temperature relief in places. Chapter 6 covers aspects of adaptation to changes that have become inevitable already. Chapter 7 outlines aspects of societal reform—it addresses the question "how can we push the necessary changes?" Finally, Chapter 8 pulls it all together.

A point of clarification: wherever I mention dollar values in this book, this refers to US dollars. If the sources report different currencies, then I have translated them using 2019 exchange rates.

## Note

1. I liken the delays in the slow processes of ocean warming and ice-sheet responses to what happens when a heavily laden freight train is called into action. A lot of energy is needed to get any movement into the train; at first, nothing seems to be happening while the engines are roaring. Then, the train begins to move and slowly, carriage by carriage, gathers pace. Eventually, the train reaches the desired speed. The engines can now be throttled down to the point where their output

needs only to overcome the wheels' rolling resistance and the wind resistance on the train. In the climate system, the engine is the radiative disequilibrium. The climate engine keeps accelerating more and more as emissions continue to grow. In consequence, the "climate train" keeps gathering speed. Now consider how difficult it is to stop a heavy freight train that is steadily moving along, let alone one that is accelerating madly.

Although imperfect, this analogy does get the message across. With our roughly 200 years of continually increasing climate forcing that is gunning the climate engine harder and harder, we have managed to awaken the slow components in the climate system: the oceans and ice sheets. These are now visibly responding to the disequilibrium, but they are lagging behind because of their inertia. If we keep gunning the engine by not reducing our emissions, or by only partially reducing them, then the climate freight train will keep gathering pace. If we stop the increase in climate forcing and keep it at its current level—by achieving zero emissions—then the climate juggernaut will keep rolling on for the thousands upon thousands of years that are needed for natural processes to reduce the forcing we have applied already. And even if we stomp hard on the brakes—by reducing emissions to zero and in addition artificially reversing climate forcing (notably by net greenhouse gas removal)—then the climate juggernaut will gradually slow down to a stop at some earlier time. But for its slowest components, this will still take a very long time; at least several centuries.

# Key Sources and Further Reading

*Call to declare a climate emergency*. Accessed April 28, 2020. https://climateemergency declaration.org

Carbon Brief Staff. *Paris 2015: tracking country climate pledges.* 2017. https://www. carbonbrief.org/paris-2015-tracking-country-climate-pledges

Casella, C. Giant gaping void emerges in Siberia, the latest in a dramatic ongoing phenomenon. *Science Alert*, September 2, 2020. https://www.sciencealert.com/ another-giant-gaping-crater-was-suddenly-found-in-siberia-the-largest-in-recent-years

Ceballos, G., et al. Accelerated modern human-induced species losses: entering the sixth mass extinction. *Science Advances*, *1*, e1400253, 2015. https://advances. sciencemag.org/content/1/5/e1400253

Hansen, J., et al. Young people's burden: requirement of negative $CO_2$ emissions. *Earth System Dynamics*, *8*, 577–616, 2017. https://www.earth-syst-dynam.net/8/ 577/2017/esd-8-577-2017.pdf

Intergovernmental Panel on Climate Change (IPCC). *Assessment report.* Accessed April 28, 2020. https://www.ipcc.ch/assessment-report/

Intergovernmental Science-Policy Platform on Biodiversity and Ecosystem Services (IPBES). *Global assessment report on biodiversity and ecosystem services.* May, 2019. https://www.ipbes.net/news/Media-Release-Global-Assessment

Kelley, C.P., et al. Climate change in the Fertile Crescent and implications of the recent Syrian drought. *Proceedings of the National Academy of Sciences of the USA, 112,* 3241–3246, 2015. https://www.pnas.org/content/112/11/3241

Long, J., and Shepherd, J.G. The strategic value of geoengineering research. *Global Environmental Change, Handbook of Global Environmental Pollution, 1,* 757–770, 2014.

McGrath, M. Climate change: warming world will be "devastating" for frozen peatlands. *BBC News,* August 10, 2020. https://www.bbc.com/news/science-environment-53726487

Meyer, R. Does climate change cause more war? *The Atlantic,* February 12, 2018. https://www.theatlantic.com/science/archive/2018/02/does-climate-change-cause-more-war/553040/

Thunberg, G. *No one is too small to make a difference.* London, UK, Penguin Random House, 68 pp., 2019. https://www.penguin.co.uk/books/315/315787/no-one-is-too-small-to-make-a-difference/9780141991740.html

United Nations. *Climate reports.* Accessed April 28, 2020. https://www.un.org/en/climatechange/reports.shtml

United Nations Climate Change. *The Paris Agreement.* Accessed April 28, 2020. https://unfccc.int/process-and-meetings/the-paris-agreement/the-paris-agreement

Vidal, J. Is climate change responsible for the conflicts we're seeing around the world today? *ENSIA,* October 15, 2019. https://ensia.com/features/climate-change-conflict-violence-extremism-draught-flood/

## Books Used for Wider Context Throughout This Book

Goreau, T.J., et al. (eds.) *Geotherapy: innovative methods of soil fertility restoration, carbon sequestration, and reversing $CO_2$ increase.* CRC Press, Boca Raton, USA, 630 pp., 2015.

Hawken, P. (ed.) *Drawdown: the most comprehensive plan ever proposed to roll back global warming.* Penguin Random House, London, UK, 240 pp., 2017.

Morton, O. *The planet remade: how geoengineering could change the world.* Granta Publications, London, UK, 428 pp., 2015.

Ohlson, K. *The soil will save us: how scientists, farmers, and foodies are healing the soil to save the planet.* Rodale Books, New York, USA, 242 pp., 2014.

Rohling, E.J. *The climate question: natural cycles, human impact, future outlook.* Oxford University Press, Oxford, UK, 162 pp., 2019.

Rohling, E.J. *The oceans: a deep history.* Princeton University Press, Princeton, USA, 272 pp., 2017.

Sachs, J. *The age of sustainable development.* New York: Columbia University Press, 544 pp., 2015.

Savory, A. *The grazing revolution: a radical plan to save the Earth.* TED Conferences, 107 pp., 2013.

Schwartz, J. *Cows save the planet: and other improbable ways of restoring soil to heal the Earth.* Chelsea Green Publishing Company, Chelsea, USA, 220 pp., 2013.

Toensmeier, E. *The carbon farming solution: a global toolkit of perennial crops and regenerative agriculture practices for climate change mitigation and food security.* Chelsea Green Publishing, Chelsea, USA, 512 pp., 2016.

# 2
# The Problem

The Human-caused Climate Crisis

## 2.1. A Look Inside the Greenhouse

To frame the problem, a brief overview is needed of Earth's energy balance and the greenhouse effect to ensure that system operation is understood in basic but essential form. For a more extensive synthesis, see Rohling (2019).

The Sun is the energy engine for Earth's climate. Heat escaping from within the Earth, including through volcanoes, is negligible in comparison. Sunlight reaches Earth as so-called Incoming Short-Wave Radiation (ISWR). At the top of the atmosphere, averaged over the year and over the entire global surface area, some 340 Watts of ISWR are received per square meter ($W/m^2$). Some 100 $W/m^2$ are bounced directly back out into space: 70 $W/m^2$ because of reflection by clouds and another 30 $W/m^2$ by reflection from Earth's surface. The remaining 240 $W/m^2$ or so are absorbed by Earth's surface (Figure 2.1).

Absorption of ISWR causes Earth's surface temperature to rise. Warm objects radiate long-wave radiation (thermal infrared). This radiation increases with temperature until a balance is reached: absorbed ISWR causes temperature rise to a point where the amount of outgoing long-wave radiation (OLWR) into space balances the amount of absorbed ISWR. In other words, for 240 $W/m^2$ of absorbed ISWR, a temperature would be reached for which OLWR is 240 $W/m^2$ as well. That temperature can be calculated with the Stefan-Boltzmann Law, and is about −18°C. But the global average surface temperature of Earth is about +15°C. What's going on? Why is Earth 33°C too warm?

At +15°C, Earth's surface radiates about 390 $W/m^2$ of OLWR out toward space. But if that was all lost into space, then Earth would be suffering a massive net energy loss, given that it gains only 240 $W/m^2$ from absorbed ISWR. In that case, Earth would be on a mad dash into a completely deep-frozen

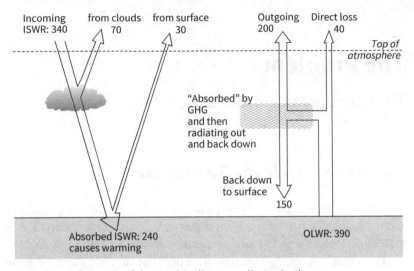

**Figure 2.1** Schematic of the Earth's climate radiative budget.

*Source*: Rohling, E.J. *The climate question: natural cycles, human impact, future outlook.* 162 pp., Oxford University Press, New York, 2019.

state, and it isn't. Something else is happening; something between Earth's surface and space is absorbing the 390 W/m$^2$ of OLWR, retaining 150 W/m$^2$ of it, and letting only 240 W/m$^2$ escape into space (Figure 2.1). That something is our atmosphere and—more specifically—the greenhouse gases in our atmosphere.

Greenhouse gases (often abbreviated in the literature as GHGs) can be identified by their spectral properties. All molecules absorb energy at specific wavelengths, in accordance with their structure and the elements they consist of. As early as the 1860s, John Tyndall performed experiments that demonstrated radiant heat absorption by different gases in the atmosphere. Though Tyndall's overall work laid key foundations to modern climate science, his experimental work on heat absorption by gases was preceded by similar work by Eunice Foote in 1856. This made Eunice Foote the first to recognize that certain atmospheric gases, notably carbon dioxide (CO$_2$), absorb solar radiation and re-radiate heat. Unfortunately, her work was almost lost to posterity because of her gender (McNeill, 2016).

Nowadays, we measure energy wavelengths with spectrometers in satellites above the atmosphere, and absorbed wavelengths show up as dark bands in the measured spectra. Such measurements have revealed that OLWR absorption in the atmosphere is dominated by water vapor (about 50%), then $CO_2$ (about 20%), and the remainder by methane ($CH_4$), along with nitrous oxide ($N_2O$), ozone ($O_3$), and a range of complex, almost exclusively industrially produced compounds.

While water vapor dominates the greenhouse effect in our atmosphere, its behavior is rather special. The atmospheric concentration of water vapor depends on temperature, through the so-called Clausius-Clapeyron relationship. Because of this, water vapor is not a "driving" greenhouse factor, but merely a compound that provides critical positive feedbacks to changes in the climate state. For a bit more detail on feedbacks, see Appendix 1. If temperature changes for whatever reason, then the water vapor content changes almost immediately in response, and by doing so it more or less doubles the initial temperature disturbance. Climate change does not start with water-vapor change. Something else, notably change in the other greenhouse gases or in absorbed ISWR, has to start the cycle. In this section, we look at the main greenhouse gases and what controls them in the natural climate state, without human interference.

The dominant greenhouse gas is $CO_2$. In the natural climate state, before human disturbances, $CO_2$ variations resulted from a vast array of intricately interlinked changes in the carbon cycle. The carbon cycle involves processes in the biological domain (biosphere), the realm of fluid water on Earth (hydrosphere) including its complex inorganic carbon cycle or carbonate chemistry, and the realm of rocks, weathering, and outgassing (lithosphere). $CO_2$ variations have been closely linked to temperature for the past 550 million years. In the past 5 million years, they have been fluctuating closely with ice-age cycles, between more-or-less present-day values of 400–450 parts per million by volume (ppmv, often shortened to ppm) in the warmest times between 3 and 5 million years ago, and about 180 ppm during severe ice ages. Over that time, climate variations were caused by small changes in the ISWR distribution over Earth's surface that were related to Earth's orbital configuration relative to the Sun. Alone, those changes would have been too weak, but they were substantially amplified (up to 5 times in polar regions) by powerful feedback processes.

The main feedbacks were $CO_2$ variations that affected OWLR retention/ loss and global changes in ice-sheet extent that affected the amount of ISWR reflection. So, in these natural cycles, $CO_2$ acted predominantly as a critical feedback to initial orbital changes in the ISWR distribution. Each doubling of the atmospheric $CO_2$ concentration adds another globally and annually averaged 3.7 $W/m^2$ of extra OLWR retention to Earth's radiation balance. This is because the radiative impact of $CO_2$ is related via a natural logarithmic function to $CO_2$ concentration changes. $CO_2$ has a long atmospheric lifetime of at least several centuries. This lifetime stands for the amount of time for a small anthropogenic addition to be brought back toward its natural level by natural processes. For comparison, the atmospheric lifetime of water vapor is 9 days.

Like $CO_2$, methane ($CH_4$) is part of the global carbon cycle, but it is dominated by biological processes. Methane is formed through maturation of hydrocarbons (petroleum) deep in the Earth's crust and by bacteria in swamps and wetlands. In the presence of free oxygen in our atmosphere and oceans, methane readily oxidizes to $CO_2$ within a matter of days; it has an atmospheric lifetime of about a decade. Ever since Earth's surface environment became replete with oxygen following the major oxygenation phase more than 2 billion years ago, atmospheric and oceanic methane concentrations have been low. In the atmosphere, they are measured in parts per billion (ppb). We don't have good records of past methane concentrations except from ice cores, which extend back in time to 800,000 years ago. But within those last 800,000 years, the ice-core records reveal variations in methane and other atmospheric gases in exquisite detail. Thus, we know that methane concentrations fluctuated between 300 ppb in ice ages and 800 ppb in the intervening warmer periods (interglacials).

Although methane fluctuations span much lower concentrations than those of $CO_2$, they cannot be ignored because, molecule for molecule, methane has a much higher global warming potential than $CO_2$. It is some 100 times more powerful than $CO_2$ over a 5-year period, some 70 times more powerful over a 20-year period, and some 20 times more powerful over a 100-year period. The drop in this global warming potential reflects the relatively rapid decay of methane in the atmosphere. During climate variations prior to human impacts, methane concentrations were especially linked to the geographic extent and intensity of tropical wetlands: organic matter decay in anoxic waters generates much methane.

At this stage, I need to emphasize one further component of control on $CO_2$ and $CH_4$, in addition to those mentioned from the biosphere, hydrosphere, and lithosphere. Although the planet's domain of ice (the so-called cryosphere) is not directly involved in the carbon cycle, it does interact with carbon-cycle processes in the other "-spheres." The cryosphere affects $CO_2$ and $CH_4$ especially via the decay of permafrost and so-called methane hydrates. Methane hydrates are crystalline cage-like structures of molecules similar to water-ice, which are stabilized by trapped molecules of methane. They are stable at elevated pressures and temperatures above freezing (0°C) at specific depths within sediments both in the ocean and on land. Decay of permafrost and methane hydrates releases masses of $CO_2$ and $CH_4$ from biological or lithological processes that had become trapped over time within these compounds, or that were sealed underneath them. Thus, increasing ocean or land temperatures can trigger a substantial feedback by which more $CO_2$ and $CH_4$ is released rapidly from such systems, exacerbating the warming. Today, there are worrying signs of accelerating, widespread permafrost and methane-hydrate decay.

In the natural climate state, nitrous oxide ($N_2O$) is emitted during much the same processes as methane. Nitrous oxide is a very powerful greenhouse gas. Where, averaged over the various periods of impact discussed earlier, 1 molecule of methane has the same global warming potential as 34 molecules of carbon dioxide, this value is about 300 for $N_2O$. Adding to its detrimental impact, it also has a fairly long atmospheric lifetime of about 110 years. Ice-core records show that $N_2O$ fluctuated between about 210 ppb during ice ages and about 270 ppb during intervening interglacials.

Ozone ($O_3$) is a gas that most of us know from its stratospheric role, where it blocks harmful incoming ultraviolet (UV) radiation. Many have also heard about surface-level ozone in smog, which is very harmful for our lungs. But not many people are familiar with the mid-altitude impact of so-called upper tropospheric[1] ozone. There, it acts as a very powerful greenhouse gas, which has a global warming potential that is about 3000 times higher than $CO_2$. But there is not much tropospheric ozone (perhaps 50 ppb), and it's very hard to keep track of because it is highly reactive and thus has a very short atmospheric lifetime. Ozone in the troposphere is closely connected to methane emissions, one of the most important so-called ozone precursors, along with nitrogen oxides. Photochemical breakdown of ozone results in hydroxyl (OH) radicals that help break down some greenhouse gases such as methane. Overall, ozone chemistry and measurements are complicated greatly by its very reactive nature. This also means no ice-core records exist for ozone. Good records for the trend in tropospheric ozone concentrations are still some way off, if they can ever be obtained at all for such a reactive, short-lived compound with large geographic and time-dependent variations.

## 2.2.  Ingredients of the Crisis

In this section, I outline the root causes of the problem that has grown during the industrial age and key controls in the climate system that determine its responses. For clarity, I refrain from listing any details beyond the level needed to understand the next chapters.

Since the onset of the industrial revolution, and especially in the last six decades, humanity has caused massive $CO_2$ release from fossil-fuel burning, cement manufacture, and net deforestation (Figure 2.2). The amount that has been released is some 430 billion tons of carbon (equivalent to 1577 billion tons of $CO_2$). We know this from four independent ways of measuring it: economical records, atmospheric $CO_2$ and ocean acidification measurements, stable carbon isotope records, and radiocarbon records. For a graphical illustration, this is as much carbon as a pure graphite column of 520 m$^2$, or almost 23 meters by 23 meters, all the way from Earth to the Moon. About one third of these emissions has gone into the ocean, where they cause acidification. In the atmosphere, the emissions of the industrial age have driven $CO_2$ values up from 272 ppm before the industrial revolution to about 410 ppm today. This represents a net additional greenhouse forcing of 1.7 W/m$^2$.

The vast bulk of $CO_2$ emissions has been driven by fossil-fuel burning, including steel production, and the wider petrochemical industry. In terms of emissions per year over the last decade or so, coal is the main culprit, with oil at almost similar levels, and gas a distant third (Figure 2.2). At an equal energy production of 1 million kJ (1 kJ is 1000 Joules), coal produces about 100 kilograms of $CO_2$, liquid oil-based fuels about 75 kilograms of $CO_2$, and gas-based fuels about 60 kilograms of $CO_2$. This underpins the often voiced need to switch from coal to gas because that would cut $CO_2$ production in half for the same energy production. However, note that while gas may produce less $CO_2$ than coal for the same energy, it still produces a lot of $CO_2$. When planning new facilities with a long lifetime, it therefore makes more sense to go for alternative energy sources that emit little to no $CO_2$.

About 8% of global $CO_2$ emissions arises from production of cement, the most important component of concrete. Concrete is the second most

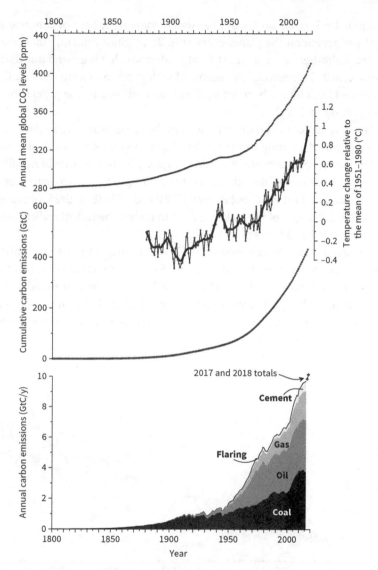

**Figure 2.2** Overview of trends in carbon emissions and global mean temperature. From top to bottom: annual mean atmospheric $CO_2$ concentrations; the global-mean annual-mean surface temperature anomaly relative to the mean of the period 1951–1980; cumulative carbon emissions in billions of tons (Gigatons) of carbon; and total annual human-caused carbon emissions from different sectors.

*Source*: Compiled and updated from Rohling, E.J. *The climate question.* 162 pp., Oxford University Press, New York, 2019. Drawing on Hansen et al. (2013, 2017) with additional information from Global Carbon Project (2018), Le Quéré et al. (2018), and BP (2019).

consumed substance on Earth after water; some three tons of concrete are used per person on the planet every year. To produce cement, limestone is heated with additions in a kiln at 1400°C, after which it is ground into a substance called clinker, which is combined with gypsum to form cement. $CO_2$ is released both by the burning of limestone itself and by the generation of energy to do so.

The dominance of carbon emissions in the global warming signal is well illustrated by the strong linear relationship between total, so-called cumulative, carbon emissions and warming (Figure 2.3). But we should not allow this to take our eye off the other causes of change in the radiative forcing of climate; that is, the difference between ISWR and OLWR. Addressing carbon emissions is but one of the tasks we face. In reality, the radiative disequilibrium needs to be addressed as a whole.

We saw before that atmospheric $CO_2$ is measured in parts per million, and that the other greenhouse gases are found in much lower quantities, measured in parts per billion. They are still very important because they have much higher global warming potentials than $CO_2$. The key lower-concentration greenhouse gases are methane ($CH_4$) and nitrous oxide

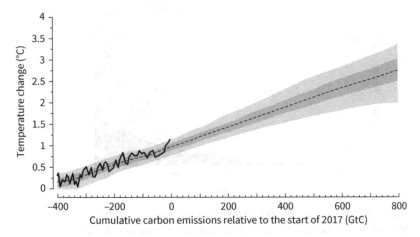

**Figure 2.3** The relationship between cumulative carbon emissions and global-mean annual-mean surface temperature.

*Source*: Rohling, E.J. *The climate question*. 162 pp., Oxford University Press, New York, 2019. Based on an original in Goodwin et al. (2018).

($N_2O$). For human impact, we also need to consider a range of synthetic greenhouse gases that arise from a variety of industrial processes that exist in even lower concentrations, but that in some cases are many times more powerful.

Methane concentrations have shot up from natural values of about 800 ppb to 1870 ppb today because of fossil-fuel exploration and human agricultural activities such as cattle raising and water-logged rice fields. Another contribution comes from decomposition of organic matter in society's solid waste landfills. In terms of amounts of carbon, this methane rise of 1000 ppb (= 1 ppm) is not a big contribution compared with the $CO_2$ rise (138 ppm). After all, once all that methane is oxidized to $CO_2$, it adds less than one percent to the $CO_2$ rise. But before it is oxidized, the methane contribution is a very important greenhouse factor because of its powerful global warming potential relative to $CO_2$. The methane increase since the industrial revolution is responsible for almost 1.0 $W/m^2$ of net additional greenhouse forcing.

In the industrial age, nitrous oxide is produced not only during the same processes as methane but also during the combustion of fossil fuels and solid waste and the treatment of wastewaters. Moreover, human activities have caused the rise in nitrous oxide especially through the use of nitrogen-based fertilizers in agriculture. Since the industrial revolution, and notably in the past century, its concentrations have shot upward to about 330 ppb. This has added 0.2 $W/m^2$ of net additional greenhouse forcing.

Finally, we come to a broad suite of exotically named, almost exclusively industrially produced, gases. These include hydrofluorocarbons (HFCs), perfluorocarbons (PFCs), sulfur hexafluoride ($SF_6$), nitrogen trifluoride ($NF_3$), a range of chlorofluorocarbons (CFCs), and trifluoromethyl sulfur pentafluoride ($SF_5CF_3$). Some have short atmospheric lifespans of a few years, but PFCs and $SF_6$ have lifetimes of 1000 years or more because there are no natural processes to break them down. Fluorinated gases are sometimes used as substitutes for outlawed substances that were found to deplete stratospheric ozone, allowing more harmful UV radiation through to Earth's surface. Unfortunately, this has proven to be a strategy of fighting one evil with another, given that the fluorinated gases are extraordinarily potent greenhouse gases. $SF_6$ has a 100-year global warming potential that is 23,500 times more intense than that of $CO_2$, with $NF_3$ scoring about 16,000, and $SF_5CF_3$ about 17,500.

This synthetic group of compounds has provided about 0.2 $W/m^2$ of net additional greenhouse forcing.

Finally, adding a further 0.2 $W/m^2$ contribution from short-lived greenhouses gases such as carbon monoxide and nitrogen oxides other than $N_2O$ that are known as $NO_x$, the combined industrial-age increases of greenhouse gases other than $CO_2$ or $CH_4$ have added an additional 0.6 $W/m^2$ or so of greenhouse forcing, so that the total of greenhouses gases has added about 3.3 $W/m^2$. But human impact on the climate system is not just about greenhouse gases. Other important impacts have also partly counteracted the greenhouse gas influence.

First, humanity has changed Earth's surface reflectivity to incoming sunlight (that is, ISWR) by changing types of vegetation cover relative to the preceding natural state. Forests are less reflective than agricultural land and even less than regions that have morphed into steppe or desert because of human actions over time. This change toward more reflective conditions has reduced net radiative forcing by about 0.2 $W/m^2$ since the year 1750, driving a slight cooling effect.

Second, our emissions concern not only greenhouse gases but also include sulfates, especially from combustion of coal, and black carbon (soot). The impacts have been complex. Black carbon is familiar from images of smoking exhausts and pollution-blackened buildings. Black carbon reduces surface reflectivity, especially when it contaminates snow or ice, or desert sand. This has contributed to warming. Sulfates have two main impacts: they cause so-called acid rain, and sulfate droplets in the atmosphere (aerosols) act as nucleation particles for the formation of bright, reflective clouds. However, not all human-caused aerosol additions are sulfates. They also include organic compounds and mineral dust. In addition to the cloud nucleation impact, aerosol particles themselves cause reflection and absorption of radiation.

By 2011, black carbon emissions had caused an increase by about 0.7 $W/m^2$ since 1750, and aerosols and their cloud effects a combined decrease of about 1.5 $W/m^2$. This together implies a net 0.8 $W/m^2$ reduction of radiative forcing by these processes. There is a critical bit of information here for the future: because of the last few decades' drive toward cleaning up air pollution, the emission of aerosols of combustion byproducts is decreasing. We are thus reducing the net negative radiative forcing of these components and consequently their net cooling influence. In consequence, the clean-up of air pollution, including the transition from combustion to renewable energy sources, is causing a decrease in this dampening effect on global-average surface warming. We could therefore expect a step up in warming even in the absence of any further greenhouse gas increases. This global-average temperature increase in response to combustion-related aerosol reduction is estimated to amount to 0.5°C, but it will not be equally distributed. North America and Northeast Asia are set to be disproportionately affected, with warming of up to 2°C (Lelieveld et al., 2019).

Overall from 1750 to 2011, these climate forcing factors yielded a total, net increase in the radiative forcing of climate of 2.3 $W/m^2$, and about another 0.3 $W/m^2$ has been added since 2011. By 2019, we've therefore reached a total increase of some 2.6 $W/m^2$ in the radiative forcing of climate. The ongoing reduction of particulate air pollution will add a further, unevenly distributed, 0.5 $W/m^2$ or so.

* * *

Following our discussion of greenhouse gases and their impacts on OLWR retention, and of changes in atmospheric and land-surface reflectivity that affect ISWR, it's time for a closer look at processes that affect surface absorption of ISWR. We previously discussed this absorption as if it is the same wherever it happens. This is not the case.

Absorption of ISWR by land surface affects only a thin surface layer. Try it out: on a hot sunny day, lift a thick paving slab (mind your fingers), and measure the temperature of the ground below it relative to that of the exposed surface of the slab. There will be a significant difference. Also, after a hot day you can feel the pavement radiate heat in the evening, but by the time the Sun rises again the pavement will have cooled to a temperature very close to that of the ambient air. Technically, we capture this by saying that land surface has a very low thermal capacity. As a result of this low thermal capacity, there are large temperature contrasts between night and day in continental settings and even larger seasonal and geographical contrasts.

It's worthwhile to pause briefly and think about ground with permafrost. Permafrost occurs over almost 25% of the northern hemisphere land-surface area, and in the southern hemisphere virtually all of Antarctica is thought to contain permafrost below the ice sheet. In terms of variability, therefore, the northern hemisphere permafrost is most sensitive because permafrost in Antarctica is protected by the ice sheet. It takes heat, and thus time, to thaw permafrost out, and refreezing takes time too. In other words, we should expect that permafrost has a higher thermal capacity than unfrozen ground. Regardless, observations show that, at the depth from the surface where seasonal to inter-annual variations are averaged out by these effects (often within 10 meters), upper permafrost temperature changes closely track regional decadal trends of atmospheric temperature. Overall, therefore, we find that the thermal capacity effect of permafrost is limited and that it does not much affect the surface or the air immediately above it. We will in the following treat land with permafrost as common land surface.

As for the oceans, they are a completely different kettle of fish. Some 93% of the energy absorbed by Earth ends up warming the vast and relatively well-mixed (over many centuries) volume of ocean water, which has very high thermal capacity. As a result, ocean temperature adjusts only very slowly. In fact, ocean temperature change to date is measurable in a straightforward manner only in the top 700–1000 meters. In deeper waters—and the oceans have a global average depth of about 3700 meters—the temperature changes are very small and require remarkable measurement precision, which has only been available on a global scale for the last decade or two. One might say that these very deep layers are only just beginning to wake up to the fact that climate change is happening. In technical terms, we talk about inertia in the system, which causes delayed responses to a perturbation.

The ongoing, slow warming of ocean waters represents a net flow of heat into the oceans, which is estimated at up to 0.8 $W/m^2$. This happens because the oceans' thermal inertia keeps ocean temperature too low for OLWR (whose intensity depends on temperature) to offset the radiative energy input into the ocean from ISWR absorption. In consequence, ocean warming will slowly but inexorably continue until ocean temperatures have risen sufficiently to restore that balance; that is, until the amounts of ISWR into the oceans and OLWR from the oceans are very similar again.

A profound conclusion can be drawn. Even if we could keep greenhouse gas concentrations completely stationary from now on, by instantly reducing emissions to zero, then surface ocean thermal inertia means that temperature will still continue to rise for hundreds of years, until it has reached equilibrium with the already-imposed anthropogenic forcing of climate. Deep ocean temperature adjustment draws that process out to a millennium or more. And if we were to try and cool the oceans down, or succeed in lowering greenhouse gas concentrations, the oceans' thermal inertia means that ocean temperatures will only adjust to *those* changes over very long timescales as well. In short, brutal terms: ocean warming is with us to stay for the next few centuries at least. All we can do is try to limit how much warmer the waters will get eventually.

Now it gets interesting: because the oceans cover some 70% of the planet's surface, and land only some 30%, only 30% of Earth's surface temperature has fully adjusted to the anthropogenic forcing of climate. Some 70% of Earth's surface has temperatures that are significantly lagging behind and that will continue to increase for centuries even if we stop all anthropogenic forcing of climate right now. (Note that this is for just surface waters; if we also consider deep waters, the timescale lengthens to a millennium or more.) Because global warming is measured as the global mean temperature increase—averaged over both ocean and land surfaces—this means that global warming will also continue to increase over the next few centuries, even if we were to stop all anthropogenic climate forcing right now. We truly have awoken an angry beast, to paraphrase the late pioneering climate scientist Wally Broecker. It may have taken some effort to awaken this angry beast because it's a bit of a laggard, but now we get to live with it—this beast won't be going back to sleep any time soon.

And then, it gets worse. It really does. This is where we turn attention to Earth's two surviving major ice sheets. They are an everyday reminder that we're still living in a period of ice ages. During most of geological time, Earth did not have such ice sheets because it was in a greenhouse state.[2] More specifically, Earth was in a greenhouse state when $CO_2$ concentrations were 700 ppm or higher. So, there is a serious risk of pushing ourselves at least partially in that direction. Today, a relatively small ice sheet occupies Greenland, and a whopping big one covers Antarctica. The Greenland ice sheet contains enough ice to raise global sea level by about 6 or 7 meters if it completely melted. The Antarctic ice sheet contains some 58 meters worth of sea-level rise. These ice sheets also respond to anthropogenic climate change. They melt, as you would guess for ice, when it's getting warmer. The 700 ppm or so is not a hard threshold above which the ice is suddenly gone. More sensitive parts respond earlier, and less sensitive parts go last. One of the most effective ways of reducing big ice sheets is by warming the oceans, which cause under-ice melting at the ice-sheet margins. This is especially effective when accompanied by air warming, which causes melting from the top. With ocean warming on a long-term trajectory with no end in sight, this does not spell a happy end for the ice sheets. Why? Because warmer oceans themselves eat into the ice-sheet margins more effectively, and also because warmer oceans carry less sea ice, and sea-ice reduction in turn allows ocean warmth to more effectively warm up the polar atmosphere. As a result, the ice sheets are facing a double onslaught.

Even when they go into full melt-back, ice sheets don't disappear overnight. The fastest rates of global sea-level rise we know of were 5 meters per century some 14,500 years ago during the last ice-age termination. This was when currently non-existent ice sheets over North America and Eurasia disappeared. One of the big questions in climate science is what sort of rates of decay we may expect in future from the great ice sheets that still exist on Earth. Many researchers agree that over the next century, it will likely be of the order of 1-meter sea-level rise, but there is significant concern that sea-level rise may accelerate far beyond that in subsequent centuries.

Another major question is to what extent the ice sheets might reduce. That is, how much sea level will rise for the current state of anthropogenic climate forcing—that is, assuming we stop any anthropogenic influence right now. This is a more difficult one to answer, but current best estimates from comparison with the most recent times during which $CO_2$ levels were similar to the present (5–3 million years ago) are that it will be at least 9 meters, converging on a best estimate of roughly 20 meters. Given the rates of rise mentioned earlier, reaching this will take many centuries. In short, the ice sheets are at least as slow, and quite possibly even slower, to adjust than the oceans.

Overall, the delayed ocean and ice-sheet responses are a signal of inertia within Earth's climate system. The very worst we can do is to continue forcing the system further out of equilibrium. The very least we can do is accept that it won't be easy to reverse the current trend of warming oceans and reducing ice sheets because we'd then be battling the same massive inertia in the opposite direction. In other words, we will be paying for a long time for the anthropogenic climate forcing we have caused already up to the present because some of the slow processes that have been set in motion will at least partially play out regardless of what we undertake.

So, what are we to do? It's pretty obvious, really. We have to (1) stop annually increasing our forcing, then (2) reduce it to zero, and eventually (3) reverse it, in a bid to remedy fast-responding elements of the system, which include the all-important ecosystems. In parallel, we need to overcome widespread belief that we might reverse *all* of human-caused climate change. Instead, we need a realization that we have already initiated changes in the system's slow-responding elements, which can no longer be stopped on timescales of less than a few centuries. Hence, we should (4) identify and prepare for adaptation to these inevitable changes. A key example is sea-level rise for centuries to come; limiting climate forcing to zero and reversing it may somewhat influence how high sea level will rise eventually, but rise it will. Our actions so far have made sure of that already. So, we need to figure out how high it may get and prepare realistic adaptation scenarios.

## 2.3. Tipping Points

Here, we need to briefly discuss so-called non-linearity in climate responses to forcing. This means that responses don't always scale directly with the amount of forcing that is applied, and this complicates the understanding of the system's predictability. Instead, systems can be slowly but inexorably pushed toward a so-called tipping point, during which time only gradual and proportional responses are seen. Then, when the system is pushed beyond the tipping point, abrupt and large, disproportional, adjustment takes place into a totally different regime. Note that this definition of a tipping point is somewhat limited in scope, but it suffices for the purposes of this book; a more complete, formal definition may be found in Lenton et al. (2008). Within our more limited scope, let's consider in a bit more depth how a tipping point works.

Tipping points mark the moment a system "tips" irreversibly from one state into another. Imagine it as the movement of a ball on a vibrating wavy surface. The ball rocks restlessly up and down the sides of one of the troughs as the wavy surface vibrates. The vibrations in this example stand for noise within the climate system, such as weather, volcanic events, seasonal cycles, and longer-term natural cycles of energy redistribution such as the El Niño Southern Oscillation or the North Atlantic Oscillation. The longer-term trend of climate change can then be represented by an ever-increasing systematic tilt of the vibrating surface. At some stage, the ball will escape out of one trough, and fall into another. This represents an abrupt change of state (trough) for the ball, caused by a systematic disturbance, such as global warming, applied to our original noisy system. As it skips from one trough into another, the ball crosses a tipping point between two states. Translated into the climate example: the climate abruptly transitions from one state to another. The climate retains its fundamentally noisy nature, but now within a different background state. Moreover, it is highly unlikely that it will cross back into its original state because of the systematic disturbance that still applies (the slope of the wavy surface). Because the responses are non-linear, simply removing the underlying disturbance would not be enough to make the system jump back into its original state; instead, a considerable extra forcing in the opposite direction would be needed. This underpins the near irreversibility of the change.

An important tipping point for climate is thought to lie between a state with considerable land-ice presence and one that is ice free. Another tipping point is thought to lie between a cool climate state with lots of methane locked away in permafrost and a warm state with no permafrost and with all the methane released into the atmosphere. Overall, some nine key subsystems within the broader climate system are suspected to be at risk of tipping. And climatologists are especially concerned that these are not necessarily independent of each other; if one flips, others are likely to follow (Lenton et al., 2019). This would lead to a cascade, or domino-effect, which can very rapidly lead to a fundamentally different climate system than the one we are used to. There is little doubt among climate scientists that such potential cascades exist. But considerable uncertainty remains about the amount of disturbance our current climate system can endure before we reach the first major tipping point.

Examples of tipping points that are especially at risk (Lenton et al., 2019) include collapse of major sectors of continental ice sheets, such as the Amundsen Sea embayment of West Antarctica, or the Wilkes Basin in East Antarctica. And the entire Greenland Ice Sheet may be doomed when global warming reaches about 1.5°C. These ice-sheet sectors may experience catastrophic instability over relatively short timescales, yet would take thousands of years to rebuild again. Clearly, avoidance is key with tipping points.

A tipping point that causes sleepless nights for lots of people concerns catastrophic methane release caused by the accelerating permafrost melt in the Arctic region. This is why NASA's 2017 Arctic Boreal Vulnerability Experiment deployed planes equipped with the Airborne Visible Infrared Imaging Spectrometer over about 30,000 square kilometers of the Arctic region. The objective was to map methane-release hotspots, where concentrations in excess of 3000 ppb were registered. They found 2 million such hotspots (Elder et al., 2020). Methane releases don't just happen on land, but also from the sea floor. As early as 2008, I took part in a geophysical study offshore from West Spitsbergen that found more than 250 methane bubble-plumes emanating from the sea floor, which we related to a regional ocean warming of 1°C within thirty years (Westbrook et al., 2009). And recent observations indicate that similar processes are beginning to act around Antarctica (Thurber et al., 2020).

On land, deforestation may cause the crossing of a tipping point especially in the Amazon rainforest. "Negative synergies between deforestation, climate change, and widespread use of fire indicate a tipping point for the Amazon system to flip to nonforest [savanna-like] ecosystems in eastern, southern, and central Amazonia at 20–25% deforestation" (Lovejoy and Nobre, 2018).

## 2.4.  A Smörgåsbord of Impacts

Understanding the processes and controls underpinning climate change is important for making better projections for the future, but they are only one side of developing adaptation and mitigation strategies. The other side concerns impacts. Global warming is a statement that is used to talk about global average temperature increase. The global average temperature increase has been convincingly demonstrated by agreement between six independent measurement compilations. One of these compilations, the Berkeley Earth compilation, was initially set up from a statistically critical viewpoint to objectively evaluate the other records. The Berkeley Earth Global Temperature Report for 2017 showed all six independent records in comparison with each other from 1850 to 2017. It is a sight to behold: the agreement between these six independent compilations is remarkable.

The global average temperature increase, however, can be slightly misleading for some people because it creates a sense that all locations on Earth should show this trend. Yet, much like the height of children in a class will vary about the average height for the class while few (if any) are exactly the average height, temperature change varies substantially from location to location and is almost nowhere equal to the global average change. In some places, there is more warming than the average, and in other places less, or even some cooling. In addition, the climate system is highly dynamic and includes a lot of temporal variability in the warming, when short-term weather systems or other internal variabilities (such as El Niño) slosh heat around between regions or between the oceans and the atmosphere. Such geographic and temporal variations do not invalidate the existence of the global average temperature increase, but merely cause fluctuations around the global average.

This cannot be emphasized enough when it comes to climate change. We should pertinently *not* expect everywhere on Earth to show the exact same trend seen in the global average. Almost everywhere, the actual, local temperature trend will differ from the global average trend because of oceanic and atmospheric processes of heat transport around the planet and influences of regional feedbacks, such as changes in snow and ice, cloud, dust, black carbon, or vegetation cover. But underneath all those regional and temporal variabilities lies the inexorable global average temperature increase that is so evident in the six independent temperature compilations mentioned in the first paragraph of this section.

Geographic differences in warming can be surprisingly large. Typically, the polar regions show much larger temperature changes than the global average because of major positive snow and ice feedbacks, as well as specific atmospheric responses. Technically, this intensified response is known as polar amplification. To date, Arctic warming has been nearly three times stronger than global average warming, and a very similar Arctic polar amplification factor is found in geological climate data for the last 100,000 years. The same geological data indicate a generally suppressed temperature response in tropical regions, at 0.5 to 0.75 times the global average. The largest temperature-related impacts of global warming are therefore found at high, polar latitudes, and the smallest impacts at low, tropical latitudes. But temperature is but one expression of climate change. The impacts of shifts in evaporation and precipitation zones, and of changing storm intensities, follow different geographic patterns than the impacts of temperature change. We will get to these a bit later after we've had a look at some of the major temperature-related impacts.

For most regional impacts, regional temperature is what matters most. The intense polar warming is associated with clear changes in sea-ice cover. The Arctic region has experienced an especially clear sea-ice decline, which is evident both in ice surface area and in ice thickness. In addition, there has been a nearly complete loss of multi-year ice that is 4 years or more old, which used to account for about a third of Arctic sea ice. Sea-ice reduction and the subsequent heat absorption by the exposed open ocean waters drives warming all around the Arctic region, including over land where it has helped to reduce the land area covered by snow by several million square miles, relative to 50 years ago. Subsequent permafrost melting, along with methane hydrate dissociation in warmer seas, is releasing methane and $CO_2$. Warmer ocean and air temperatures have also caused rapid acceleration of ice mass loss from the Greenland ice sheet. In contrast to melting sea ice, which floats on the ocean and therefore does not affect sea level, melting of continental ice sheets does cause global sea-level rise.

Things work differently around Antarctica. Here, sea-ice variations depend on multiple controls, including thermal isolation of the Antarctic continent by the surrounding Southern Ocean and interactions between powerful polar easterly winds near the continent and intense westerlies over the open Southern Ocean. This has resulted in complex sea-ice variability with an underlying weak trend of increasing sea-ice cover, overprinted by an about 5-year cycle that has intensified substantially from 1983 until 2015. Then, from 2015 to 2017, a sudden 40% decline took place, following which sea-ice cover around Antarctica has remained small. But it is too early to say whether the sharp drop represents an extreme part of the 5-year cycle or whether it was the start of a new era with small sea-ice cover around Antarctica. Several years of further observations are needed to decide this.

Given that sea ice has low salinity, its formation rejects salt into ambient waters, and relatively fresh meltwater is added to the ocean surface when sea ice melts. The amount of salt in water determines its density (together with temperature). Land-ice melting adds even fresher water to the ocean surface. Thus, changes in sea-ice and melt-back of land-ice cause ocean-water density variations, which in turn affect ocean deep-water circulation. Under conditions of net sea-ice decline and large-scale land-ice reduction, deep-water formation may become inhibited. Embayments around Antarctica, and the Nordic Seas (the basin between Norway, Svalbard, Greenland, and Iceland), are critical deep-water source areas for the world oceans. Oceanic deep-water circulation may therefore become weakened as substantial ice melt occurs. Since deep-water formation in the Nordic Seas draws warm surface waters northward in the form of the current system that comprises the Gulf Stream and North Atlantic Drift, a reduction of deep-water formation in the Nordic Seas will draw less of this warm water northward, which results in regional surface-water cooling. Climate models for the future show distinct regional cooling because of this process. This is one of those instances where global average temperature increase drives regional processes (ice melt) that in turn cause a distinct regional cooling, which bucks the global average warming trend.

Finally, loss of frozen habitats in general, including sea ice, has massive impacts on ecosystems comprising species that have evolved over hundreds of thousands of years in adaptation to the particular, extreme conditions. These species will lose their competitive advantages and succumb to invasive species that are better adapted to warmer conditions. As a result, food webs will change, specialists will become marginalized, and extreme specialists will become extinct. Loss of Arctic sea ice is expected to massively alter primary (algal) productivity in the Arctic Ocean, with deep impacts on the Arctic food web. This critically endangers indigenous seals, fish, Arctic foxes, narwhals, and polar bears, along with many other indigenous species that might be less photogenic but still are crucial to healthy ecosystem operation. Lessons may be learned from the northern Bering Sea, where productivity changes caused by warming and sea-ice loss have already resulted in a dramatic change in the ecosystem, from one with diving ducks and walruses as top predators that fed on sea-floor organisms, to one dominated by pelagic fish. In Antarctic coastal regions, glacier and sea-ice changes are causing diatom algae (the preferred food of krill) to be replaced by less palatable cryptophyte algae. The resultant changes in krill stocks in turn reverberate through the wider food web, including birds, fish, squid, seals, and whales. There are also impacts of seasonal shifts; for example, near Palmer Station, changes in the sea-ice season shifted the period of maximum krill stocks away from the Adélie penguins' peak foraging season. This contributed to an 80% crash in the Adélie penguin population and invasion of Chinstrap and Gentoo penguins. The species replacement patterns differ from region to region around Antarctica, depending on which species is negatively affected and which is favored.

The global temperature increase not only causes shrinkage of polar regions, but also affects all other climate zones on Earth. For example, subtropical climate zones are expanding toward higher latitudes, which causes aridity increases at the poleward margins of these zones. Tropical regions are noticeably expanding as well. Such climate zone shifts profoundly affect the habitats of flora and fauna and are especially detrimental to organisms that are immobile and incapable of migrating with the changing conditions. Furthermore, widespread upward displacement of climate zones toward higher altitudes in mountainous regions is compressing the ecological niches of high-altitude adapted biota. Overall, some 75% of marine species have experienced about 1000-kilometer poleward shifts in their habitats, while more than 50% of land-based species have experienced poleward range shifts of as much as 600 kilometers and upward shifts of some 400 meters.

A survey of a wide variety of environments has found that between 1930 and 2010 warming-related shifts in oceanic ecosystems have driven an average 4% drop in the maximum sustainable yield of fisheries-relevant populations. That average number doesn't sound like too much, but much greater than average losses were recorded in several important regions. For example, the actual value for East Asia is 15–35%. Moreover, ocean warming—especially through increasingly frequent marine heatwaves—is placing species critical for marine biodiversity under severe stress globally, such as corals, seagrasses, and kelps. For example, corals suffer from bleaching when they are exposed to too much warming; the tolerance can be as little as 1°C. Because of ocean warming, and marine climate zone and current shifts, coral bleaching events are some five times more frequent today than they were in the early 1980s. When bleaching is infrequent, many corals typically recover over a couple of years. But when bleaching becomes too frequent, there is not enough time for recovery, and most of the reef will die. Reef monitoring indicates that we have approached this critical point globally.

A highly visible change in climate conditions is the increasing intensity and frequency of heatwaves, including regions of North America, Eurasia, China, and Australia. The European heatwave of 2003 led to an estimated heat-related death toll of more than 70,000, in stark contrast with about 3000 in regular years. The Russian heatwave of 2010 resulted in some 55,000 heat-related deaths, as well as more than 500 wildfires near Moscow and grain-harvest losses of 30%. Summer 2018 saw exceptionally widespread heatwaves across both Europe and Asia, associated with anomalously frequent and intense forest fires. The heatwave of June 2019 caused record temperatures of 45°C in France and spontaneous combustion of manure-stores in Spain. Studies have attributed the increasing heatwave frequency to global warming and also indicate that the risk of similar extreme events increases dramatically over just a few decades into the future. Australia has also seen a remarkable increase in extreme heat events over the past 120 years. In the United States, however, the trend in heatwave frequency remains dominated by natural variations.

But heatwaves are not the entire story. The general tendency toward conditions of prolonged heat brings an even more widespread threat of heat stress and heat death, especially in the humid tropics (Mora et al., 2017; Xu et al., 2020). Sustained hot conditions can prevent the body from cooling down sufficiently, to the point where it reaches a lethal core temperature of about 42°C. The body's heat loss is also influenced by air humidity because evaporation of sweat helps us cool down. We all know how the increased air humidity in muggy weather reduces the effectiveness of cooling by perspiration. Thus, lethal conditions can develop during exposure without relief at temperatures of 30–35°C at more than 90% humidity, or at temperatures of roughly 55°C at close to 0% humidity. For sufficient relief, it has been found to be particularly important that nights drop below 28°C. Other larger mammals are affected at roughly similar conditions, with the caveat that the perspiration mechanism in humans is much more efficient than in most other mammals. Many regions in the humid tropics are very close to the danger limits already, and only limited additional warming is needed to cause severe stressful to lethal conditions. The populous regions of Amazonia, central Africa, India, South-East Asia, and Indonesia are of specific concern. The vast majority of people and animals, critically including livestock, won't be able to access air conditioning for relief. Together with other climate change drivers—drought, flooding, and sea-level rise—this poses a threat of widespread mortality, collapsing food supplies, mass migrations, spreading of infectious diseases, and associated international security risks. Prevention is the best response to such overwhelming risks, and it is feasible because there is little inertia involved in the direct temperature response. In fact, projections indicate remarkable differences in the sizes of the regions exposed to heat stress and heat death for different future emission scenarios. This emphasizes the urgency to transition sooner rather than later to a low or no emission scenario.

Warming causes an increase in evaporation and a higher water-vapor transport capacity in a warmer atmosphere. This drives a notable increase in the occurrence of regional droughts where evaporation dominates and flooding where precipitation dominates. Moreover, increased cycling of water through the atmosphere has been linked with increased intensity of large storms, notably hurricanes. The powerful rainfall and winds of hurricanes are extremely destructive, and the winds in addition drive storm surges that can reach many meters. Thus, hurricanes cause extensive flood and wind damage. Projections for the future under sustained warming see a major increase in both destructive aspects of hurricanes (rainfall and winds). Major storms outside the tropics are also expected to become intensified as more water vapor, and thus energy, is cycled through the atmosphere. Hurricane and general major storm activity is greatly influenced by both surface-ocean temperature and geographic contrasts in surface-ocean temperature. Hence, storm activity is related to emissions via a system with considerable inertia (surface-ocean temperature), which likely limits the level of immediate response we might expect to emissions reduction, although substantial response might be expected over multi-decadal timescales.

This brings us to hydrological changes. The contrast between wet and dry regions and between wet and dry seasons will probably increase, with regional exceptions (Trenberth et al., 2014). This is sometimes simplified as "wet gets wetter, dry gets drier." The same study also stated that, over land, much of the extra heat of climate change goes into drying, so that (naturally triggered) drought should set in more quickly, become more intense, and last longer. It concluded that climate change may not create droughts, but it can both worsen and expand them in the subtropical dry zone—and, as we have seen previously, expand this climate zone poleward.

The East Asian Summer Monsoon is of critical importance to the livelihoods of hundreds of millions of people in the region. Records for the East Asian Summer Monsoon that span 145 years show a remarkable weakening/drying trend in that monsoon system since the 1880s. The Indian monsoon, critical to hundreds of millions of other people, appears to have suffered a similar decline, from about 1860 to the present. Relationships are being investigated between the weakening monsoons and warming of surface waters in the tropical Pacific and Indian Oceans. If this is the controlling factor, then considerable inertia would apply to the monsoon responses. As with storm intensity, this likely limits the amount of immediate response we might expect to emissions reduction, instead stretching the response over multi-decadal timescales.

Major flooding in the valleys of glacier-fed rivers is expected while glaciers retreat rapidly under warming conditions. That is, until the glaciers have disappeared, by which time their critical feed into such rivers will disappear entirely. Glaciers are receding globally and many are expected to have disappeared by the middle of this century. As an aside—to ward off a common critique—this statement is *not* invalid because a handful of glaciers is still stationary or growing. Each glacier responds to its own specific balance between mass accumulation of snow/ice and mass loss from melting and calving. So, a glacier fed by a basin with very high snow accumulation may be stationary or even grow, despite major mass loss. A good example is the Perito Moreno glacier in Patagonia (Figure 2.4). But globally, the disappearance of glaciers will have major consequences for water resources, especially where dry-season river flows depend on glacier meltwater, notably from the Himalayas and Hindu Kush, the Andes, Rocky Mountains, and European Alps. Before this lack of meltwater feed into the rivers becomes an issue, however, there is increased flooding into those valleys as the

**Figure 2.4**  The author at the terminus of one of the few remaining glaciers in the world that are not receding, the Perito Moreno glacier, Los Glaciares National Park, Patagonia (November 2017).
*Source*: Author's own photograph.

glaciers melt back more than usual. And this flooding can have disastrous consequences. Receding glaciers draw back from their terminal moraines—the materials they have bulldozed into messy ridges of sand and boulders along their edges. Meltwater collects in lakes behind these ridges, but those ridges do not have the consistency of a sound dam. While some ridges weep through and slowly disintegrate, others give way abruptly and cause catastrophic floods that sweep away everything before them in the downstream valleys.

The impacts discussed here represent just a selection. The IPCC Fifth Assessment Report contains a much broader overview and a wealth of carefully documented supporting literature. But I would be remiss if I omitted to highlight one further key impact of humanity's carbon emissions. It concerns the oceans. The oceans have absorbed not only 93% of the heat, but also about a third of all emitted carbon. Dissolution of this much carbon into the ocean water over the short timescale involved (about 200 years, but mostly the last 60 years) has activated the seriously detrimental process known as ocean acidification. Ocean pH, a measure of acidity, has dropped from about 8.2 to about 8.1 since the onset of the industrial revolution. While this may not sound like a lot, pH is a logarithmic indicator of acidity, so the 0.1 drop represents a 25% increase in acidity.

If humanity continues along a "business-as-usual" scenario of emissions, surface-ocean pH will become 7.7 or 7.8 by the end of this century. In humans, a pH reduction of only 0.2 causes serious health issues, including seizures, coma, and death. Fish have similarly low tolerances to pH change, and trying to restore its pH when it drops costs a fish so much energy that growth, mobility, and reproduction are stifled. And the impacts do not stop at fish. Small pH changes equally affect other marine organisms, whose pH like that of fish depends on the pH of their ambient waters. Key processes that are affected include growth, reproduction, carbonate skeleton formation, and capacity of food digestion in larvae. This endangers the base of the food web. As another aside, to fend off remarks that carbonates in the ocean sediments will dissolve and offset the acidification: yes, that is true, but it takes hundreds of years, up to 10,000 years—too long to provide society any relief from the profound impacts. In simple terms, sustained ocean acidification does threaten the entire marine food web, and natural processes are too slow to fix it.

## 2.5.  A Taste for Change

So, where does humanity stand on all this? What has been the international response over the last few decades? Although the principles of greenhouse forcing of climate were acknowledged as early as the late 1800s, the discussion of the threat of global warming, or global climate change, only started to gain momentum from the late 1970s. Then, in testimony to a US Senate committee on June 23, 1988, James Hansen of the NASA Goddard Institute for Space Studies laid out unambiguously that human-caused increase in the greenhouse effect was driving the warming climate. This was the start of focused research into the phenomenon on a global scale, and a first international assessment of the growing body of research was presented in the IPCC's first assessment report of 1990. Through 2019–2020, writing of the sixth assessment report has been in full swing.

The IPCC process has highlighted a deep scientific consensus about not only the general trend of warming but also the associated uncertainties, geographic variability, impacts, and options for remedial action. The fifth assessment report had a combined lead-author list of 831 experts from disciplines that span meteorology, atmospheric chemistry, physics, oceanography, statistics, engineering, ecology, social sciences, and economics. But, importantly, the general trend of greenhouse warming has also been evaluated elsewhere, by bodies that definitely cannot be accused of a green idealist agenda. Notably, petroleum giant Exxon performed its own climate change assessment as early as 1982, in the form of an internal document. It shows exceptional prescience about the development to today's $CO_2$ concentrations, and how they would further develop through this century (their Figure 3).[3] It also predicted the associated global average temperature changes with uncanny accuracy, as well as the superimposed polar amplification and its threat to polar ice sheets. Shell, another petroleum giant, produced a similarly prescient study in 1986, which was published as a confidential report in 1988. In this report, Shell accounted for its own contribution to carbon emissions, highlighted specific impacts of climate change including region-specific consequences, and advocated that the energy industry ought to consider how it can play a part in taking remedial action.

On the political side, May 9, 1992, saw the adoption of an international treaty called the United Nations Framework Convention on Climate Change (UNFCCC), whose objective was to stabilize atmospheric greenhouse gas concentrations at a level that would prevent dangerous anthropogenic interference with the climate system. Signed a month later at the Earth Summit in Rio de Janeiro, it set non-binding limits on greenhouse gas emissions and lacked an enforcement mechanism. The members, or parties, to the convention have been meeting every year since 1995 at the so-called Conference of the Parties (COP) to evaluate progress. In 1997, the Kyoto Protocol was adopted, which entered into force in 2005 and commits states to reduce greenhouse gas emissions.

The Kyoto Protocol runs over two commitment periods, the first from 2008 to 2012 and the second (the Doha Amendment) from 2012 to 2020. There were 192 parties to the Kyoto Protocol, but only 128 have ratified the second commitment period. As a result, the second commitment period has not come into force because that requires ratification by 144 parties (three quarters of the total). Meanwhile, the annual UNFCCC meetings developed new measures for the period after 2020, which eventually led to adoption of the 2015 Paris Agreement. The Paris Agreement has been ratified by 185 states and the European Union. A further 11 states have signed the agreement, but not ratified it (yet). Although the United States was withdrawn from the agreement by President Trump, it was subsequently recommitted by President Biden. The Paris Agreement aims to limit the increase in global average temperature to within a maximum of 2°C above pre-industrial levels and to pursue efforts for limiting it at 1.5°C. To meet these targets, all participating parties are required to determine, plan, and regularly report on the actions undertaken.

There is an area of concern. The Paris Agreement's target-setting process seems to overlook that the stated targets are somewhat deceptive. They are just a year-2100 snapshot view of a warming process that is continuously developing through time. Data for climate change in the recent geological past and long-term modeling indicate that 2°C warming by 2100 would be followed by further warming to roughly 3°C by 2200 and to about 4°C over a few more centuries, even if there are no additional anthropogenic emissions. This happens because of the delayed surface-ocean warming and slow feedbacks such as ice-sheet retreat, vegetation change, and permafrost methane/$CO_2$ release. Other major feedbacks may also kick in, possibly even on shorter timescales. For example, field experiments across North America, Europe, and Asia indicate that every 1°C global average soil-surface warming may result in a release from soils of 30 GtC and that continued high business-as-usual fossil-fuel emissions could drive almost 2°C soil warming and release of 55 GtC as early as 2050 (Crowther et al., 2016). This would be a significant, unwelcome source of "new" carbon input into the climate system: a 55 GtC input would cause 15–17 ppm increase in atmospheric $CO_2$ concentrations, after accounting for gas exchange with the ocean. And then there are the tipping points mentioned in the previous section. Taking all this on board, some climate researchers (myself included) consider even the Paris Agreement's most ambitious 1.5°C limit to be too generous and instead argue for a limit at 1°C by 2100 (for more detail, see Hansen et al., 2017). Clearly, there is great urgency in meeting the Paris Agreement targets and a strong preference to do even better. Let's have a look at our scorecard relative to these ambitions.

At the time of the UNFCCC in 1992, the annual average $CO_2$ concentration was 356 ppm. Five years later, at the time of adoption of the Kyoto Protocol, it had increased by an average of 1.4 ppm per year. Over the 8 years until the Kyoto Protocol entered into force, the $CO_2$ level increased by 1.9 ppm per year. From then until the intended start of the Doha Amendment in 2012, $CO_2$ rose by an average of 2.0 ppm per year. In the 3 years to the Paris Agreement, the annual average $CO_2$ concentration climbed by almost 2.2 ppm per year, and in the 4 years since then the rise reached an average of more than 2.5 ppm per year, toward an estimated average 2019 value of about 411 ppm. These numbers (see also Figure 2.2) illustrate with terrible clarity that all the agreements reached so far have failed to deliver any appreciable reduction in human-caused emissions; instead, the emissions have been increasing through time. Looking into the future, things do not look very promising either, at the moment. The Paris Agreement may have been a breakthrough event in international politics, but the subsequent pledges for action, or Intended Nationally Determined Contributions (INDCs), fall well short of delivering on the 2°C or 1.5°C targets—they are more in line with warming of about 3°C by 2100. And to make matters worse, almost none of the parties is on target with their own pledges.

A May 2019 working paper of the International Monetary Fund (IMF) has shed some light on the likely cause for society's structural failure to reduce its greenhouse gas emissions. It calculates fossil-fuel subsidies for 191 countries (Coady et al., 2019). These subsidies are defined as fuel consumption times the gap between existing and efficient prices, where efficient prices are those warranted by supply costs, environmental costs, and revenue considerations.

Globally, the fossil-fuel subsidies were calculated at an astonishing $5000 billion (a billion is a thousand million), which is more than 6% of the global Gross Domestic Product. China is the main subsidizer, at $1400 billion. The United States is second, with fossil-fuel subsidies of about $650 billion, an amount that is almost identical to the total annual US defense expenditure.

There is some debate about the total value of fossil-fuel subsidies stated in the IMF report, because the report followed an unconventional approach to the definition of subsidies (that is, it includes so-called externalities), although many argue that this approach better represents the true cost to society. In contrast to the exact total amount, proportional contributions are less debated (Table 2.1). Interestingly, underpricing for local air pollution (smog) accounts for twice as much of the post-tax subsidy as underpricing for global warming impacts. The public health agenda is providing critical impetus to the renewable energy increase in and around polluted cities. It also deserves much more attention, alongside the global warming argument, in framing a rationale for reducing fossil-fuel subsidies. After all, reduction of emissions detrimental to health represents a significant social and economic co-benefit to reduction of emissions responsible for climate change.

Alternative energies received roughly $90 billion in global subsidies in 2011, out of a Bloomberg-listed $324 billion total global investment in renewable energy that year. Simply based on those numbers, renewables appear to receive a global level of subsidy of about 28%. But the International Renewable Energy Agency reports that the actual level in the major adopters—China, North America, and western Europe—was only about 10%. The difference possibly arises from what each analysis exactly includes in the term subsidy. As long as fossil fuels remain as heavily subsidized as they are today, relative to renewables, there is no viable business case for a

**Table 2.1** Proportion of fossil fuel subsidies out of the global $5000 billion (Coady et al., 2019).

| Method | Post-tax subsidy percentage (%) |
| --- | --- |
| Coal | 44 |
| Oil | 41 |
| Gas | 10 |
| Fuel inputs into electricity | 4 |

large-scale transition. Having said that, change is afoot. The investment bank Lazard reported in November 2018 that in developed countries the unsubsidized levelized cost of energy then was $27–45 per megawatt-hour for coal from existing plants versus $29–56 for a new-built wind farm and $36–44 for a new-built solar farm.

In the same month, Carbon Tracker reported that 42% of coal plants already operate at a loss, that it already costs more to run 35% of coal power plants than to build new renewable generation, and that wind and solar will be cheaper than 96% of existing coal power as early as 2030. Major savings are possible by closing plants in line with the Paris Climate Agreement instead of continuing business as usual plans (Table 2.2). Thus, economic drivers will help to phase out energy generation from coal and other fossil fuels, although this natural tendency will be delayed unnaturally by sustained fossil-fuel subsidies.

Public awareness of the favored position of fossil fuels and the need to transition to renewables is increasing, along with awareness of concepts such as carbon footprint, carbon neutrality, emissions reduction and emissions offsets, and so on. There is also growing public concern about climate-change impacts, including sea-level rise, weather extremes (heatwaves, flooding, storms, and hurricanes), reef bleaching, and rapidly accelerating species extinction rates. The resultant public pressure propelled the political consensus behind the Paris Agreement and also the recent declarations of a climate emergency by several nations or states within nations.

Hope springs eternal, so let's assume that the enhanced awareness would miraculously trigger an abrupt and complete switch of global society to zero-emissions energy. Would that suffice to completely stop the rise of

**Table 2.2** Savings possible by closing plants in line with the Paris Climate Agreement, instead of continuing business as usual (Carbon Tracker, 2018).

| Nation | Possible savings (billion $) |
| --- | --- |
| China | 389 |
| European Union | 89 |
| United States | 78 |
| Russia | 20 |

greenhouse gas concentrations? Unfortunately, the answer is no, because lots of emissions either cannot be avoided or are very difficult to reduce. Examples are those related to the petrochemical manufacturing or cement industries, deforestation and land clearance, diffuse methane leakages from agriculture and fossil-fuel extraction (including fracking), and potential soil-carbon release from soils in warming conditions. Many of these don't cause emissions from a discrete point source, and these emissions are therefore difficult to capture at source and then process to keep them out of the atmosphere. In consequence, some rise in atmospheric greenhouse gas concentrations would continue even if society would have adopted 100% zero-emissions energy.

If all these continuing emissions could be captured and removed from the atmosphere, then total greenhouse gas rise might be halted. That is, the greenhouse gas concentrations would be kept steady at their current value; they would not appreciably drop under zero-emission conditions. Importantly, development of the necessary measures of emissions reduction and avoidance of addition of new emissions will take time. Thus, emission reduction pathways with time may be represented simply by assuming certain percentage reductions per year (Figure 2.5) or can be quantified in more elaborate future projection scenarios as used by the IPCC. Either way, projections can be made into the future, which allows assessment of how well each scenario performs relative to the Paris Agreement's warming limits, and of the potential carbon removal that each may need to rectify any excess emissions. All scenarios need to start with society's current track record, which is far from ideal. In fact, a further 10 GtC of emissions is added every single year to the total amount of carbon that needs to be dealt with (Figure 2.2). And future population growth and development will cause this annual addition to grow steadily unless rigorous emissions reduction schemes are implemented (Figure 2.5).

Figure 2.5 provides a basic example assessment along these lines. It is a straightforward thought experiment that considers past emission rates as well as the impacts of population growth and ongoing development to evaluate required levels of emissions reduction and carbon removal from the climate system. The idealized set of calculations gives sufficiently reasonable approximations to illustrate our predicament over the next century, but should not be taken as detailed predictions. Panel a shows United Nations median estimates for global population growth. Panel c shows

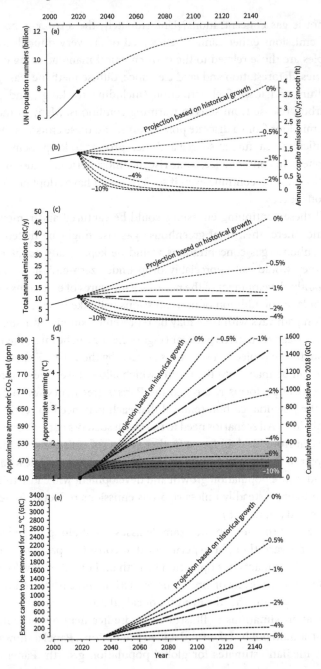

**Figure 2.5**  Idealized projection scenarios until the year 2150.

**Figure 2.5** Continued

**a.** United Nations global population estimates.

**b.** Annual per capita carbon emissions, in tC per person per year, obtained by dividing data from panel **c** by data in panel **a**.

**c.** Total global annual carbon emissions, in GtC per year, obtained (0% line) from a smooth fit through the historical data in Figure 2.2 and extrapolation of that fit function until 2150, and for the other scenarios by imposing annual percentage-wise emissions reductions starting in 2019.

**d.** Global cumulative carbon emissions from 2019 onward, obtained by adding up the successive annual values from panel **c**. Grey zones delineate the Paris Agreement 1.5°C and 2°C warming limits. Axes on the left-hand side show approximate warming levels and atmospheric $CO_2$ levels associated with the cumulative carbon emission levels shown on the right-hand axis. In reality, the relationship between $CO_2$ and temperature change is slightly non-linear.

**e.** Amounts of excess carbon that need to be removed from the climate system for each scenario that exceeds the Paris Agreement warming limit of 1.5°C in panel **d**.

The heavy dashed lines in **b-d** indicate a specific scenario that keeps global total annual carbon emissions stable from 2019, regardless of population growth and development. This is a scenario of perfect new additional carbon emissions avoidance.

global annual carbon emissions, represented by a smooth fit through the historical data in Figure 2.2. The 0% line indicates a forward projection to 2100 based on the historical growth in annual carbon emissions. The 0% line in panel **d** shows the cumulative carbon emissions from 2018 onward. The 0% line in panel **b** shows the relevant global-average per capita carbon emissions, based on the 0% line in panel **c** divided by the population number in panel **a**. Then, the other lines show alternative scenarios, for 0.5, 1, 2, 4, 6, 8, and 10% reduction in carbon emissions per year, starting in 2019. In panel **d**, there are two shaded zones. One delineates the Paris Agreement's 1.5°C warming limit and the other the 2°C limit. It is evident that only scenarios that include 7% per year or more reduction in global carbon emissions can hope to comply with the 1.5°C warming criterion. And that's assuming that these drastic emissions cuts started from 2019; in reality, they haven't. The heavy dashed lines in panels **b–d** indicate a special scenario, in which global total annual carbon emissions remain stable from 2019; this would be the case if the only measure taken was to prevent any new carbon emissions being added from continued population growth and development. In other words, in this scenario all additional, new emissions are perfectly avoided. It is obvious from the graphs that this measure alone does not suffice at all for getting us even close to the Paris Agreement's 2°C warming limit. Finally, panel e shows how much carbon needs to be removed for the various scenarios that exceed the 1.5°C limit, in order to have any hope of bringing them back within that limit. For scale, each 100 GtC is equivalent to a square column of pure graphite carbon, like an artist's pencil, of 11 meters by 11 meters, stretching all the way from Earth to the Moon.

When proper calculations are done using a wide range of more realistic emissions scenarios and radiative balance calculations that also account for non-$CO_2$ effects, it turns out that meeting the Paris Agreement's 1.5°C limit will require carbon removal of 70–280 GtC by 2100. Expressed in terms of $CO_2$, this translates into removal requirements of 260–1030 $GtCO_2$ (Rogelj et al., 2018). The low end of this range applies to the most favorable scenarios, and the high-end values apply to scenarios where business-as-usual (as in the past two decades) is allowed until 2030 or beyond. None of these scenarios accounts for surprises in the slow feedback processes that we saw before. It is assumed that the slow feedbacks will simply keep developing in step with anthropogenic changes in the same proportion they have done over the past two centuries. If the slow feedbacks throw us a surprise, such as a methane and/or $CO_2$ release from decaying permafrost at a historically unprecedented scale, then the calculated values represent underestimates of what will really be needed.

So, even without thinking about any unwelcome surprises from the slow feedback processes, there is a gargantuan challenge already in realizing 70–280 GtC removal from the climate system by 2100, along with permanently locking it away from coming back. That's an awful lot of carbon to deal with. Clearly, the first thing that's needed is to ensure that society starts tracking the low-end scenarios globally, which means extremely drastic reduction of greenhouse gas emissions, along with breaking the ever-increasing trend of total energy demand. But even then, we will need massive removal of carbon from the climate-and-ocean system, which is interlinked through gas exchange. Such removal is often captured under the term carbon dioxide removal (CDR). This leads us into the scope of the following three chapters.

Rapid emissions reduction is the cheaper option, at least when looking at the "easy" aspects that could be implemented immediately, such as energy transition to renewables. There is an increasing public groundswell toward implementation of renewable energy, with expansion of both rooftop solar and solar and wind farms accelerating on a global scale. Widespread adoption helps to push down component prices and infrastructure needs through economy of scale and upscaling of infrastructure from individual to large collective scales. These are key reasons why new wind and solar farms are becoming cheaper than running coal-fired plants, as we saw before. Emissions reduction options are the topic of Chapter 3. Chapter 3 also gives attention to another, complementary possibility, namely that society ought to decrease its total energy demand. This would be invaluable for boosting the effectiveness of a transition to renewable sources for total emissions reduction. After all, even a 60% increase in renewable energy provision could be easily offset by a 60% increase in total energy demand, which would leave the actual amount of fossil fuel combustion, and thus emissions, more-or-less constant. The optimized strategy, instead, involves a decrease or (at least) stabilization in total energy demand alongside the rapid transition to renewables.

CDR, and removal of other greenhouse gases, is more expensive than emissions reduction or avoidance and—more importantly—so far mostly available in laboratory or experimental stages only. Lack of widespread implementation means that it still remains an expensive undertaking. However, scientists and engineers worldwide are recognizing this as a critical set of technologies and methods for the future, and also as a rich nurturing ground for new concepts, ideas, and development. Many different solutions have been suggested already, and new ones are being added every day. Some with little to no evaluation of potential downsides and some with little to no attention to the issue of scale that is needed. It is true that each single viable solution will be needed to meet the challenge, but it is also true that some methods lend themselves much better to scaling up toward Gigaton scales than other methods. And if there are limited funds for development, a case might be made to focus on the large-impact solutions first. Greenhouse gas removal options are the focus of Chapter 4.

We have seen that future scenarios are still available with outcomes by 2100 that agree with the Paris Agreement targets. However, we have also seen that every year of delay and continuation of business as usual makes it more and more unlikely that enough CDR can be undertaken to keep or get warming at or below the Paris Agreement targets. This is especially true given that we don't even know yet which CDR approaches would be best suited for the gigaton-scale challenge. With this in mind, proposals have been floated about solar radiation management (SRM), which is all about managing the amount of ISWR in the energy budget of climate (Figure 2.1). These approaches therefore do not reduce ocean acidification. Some of these proposals are new, while others are revitalizing proposals that have been made in the past as full-on solutions to global climate change. Today, they are packaged more as rapid-relief solutions that will help to buy time during which CDR can be brought up to speed, which will then reduce the root-cause of climate change and alleviate ocean acidification. This class of proposals will be evaluated in Chapter 5.

# Notes

1. Earth's atmosphere comprises several layers. Most relevant to this book are the lowermost layer—the troposphere in which the weather plays out—and the stratosphere. The stratosphere is the layer of gases between on average 15 and 50 kilometers of altitude that is not affected by the surface weather. The boundary between these two, the tropopause, resides at up to 20 kilometers in the tropics and less than 10 kilometers at the poles.
2. Natural greenhouse conditions came about because of minute imbalances between $CO_2$ outgassing from volcanoes and other such processes and $CO_2$ drawdown through weathering and organic matter burial. These minute imbalances were maintained over hundreds of thousands to millions of years and resulted in long-term $CO_2$ cycles, which determined past cycles between (common) greenhouse states and (rare) ice-age states of Earth's climate. Human emissions drive $CO_2$ build-up that is typically 100 times faster than $CO_2$ build-up (or removal) by natural processes. For more detail see Wallman and Aloisi (2012) and Rohling (2019). For completeness, note that there is a mistake regarding the net carbon flux in Figure 4.1 of Rohling (2019), which I indicated to amount to a net loss of 0.4 GtC per year. This was based on a misrepresentation of the net weathering impact, which should be +0.1 GtC rather than −0.3 GtC.
3. EXXON Research and Engineering Company, Coordination and Planning Division. *$CO_2$ greenhouse effect: a technical review*. April 1, 1982. http://www.climatefiles.com/exxonmobil/1982-memo-to-exxon-management-about-co2-greenhouse-effect/ (last accessed May 26, 2021).

# Key Sources and Further Reading

Barker, S., and Ridgwell, A. Ocean acidification. *Nature Education Knowledge, 3*, 10, 21, 2012. https://www.nature.com/scitable/knowledge/library/ocean-acidification-25822734

Berkeley Earth. *Global Temperature Report for 2017* (contains a comparison graph between six independent measurement compilations). http://berkeleyearth.org/global-temperatures-2017/

Bevis, M., et al. Accelerating changes in ice mass within Greenland, and the ice sheet's sensitivity to atmospheric forcing. *Proceedings of the National Academy of Sciences of the USA, 116*, 1934–1939, 2019. https://www.pnas.org/content/116/6/1934

Biskaborn, B., et al. Permafrost is warming at a global scale. *Nature Communications, 10*, 264, 2019. https://doi.org/10.1038/s41467-018-08240-4

Block, B. A look back at James Hansen's seminal testimony on climate, part one. *GRIST Climate & Energy*, June 2008. https://grist.org/article/a-climate-hero-the-early-years/

BP. *Statistical Review of World Energy*, 68th edition. 2019. https://www.bp.com/content/dam/bp/business-sites/en/global/corporate/pdfs/energy-economics/statistical-review/bp-stats-review-2019-full-report.pdf

Carbon Brief Staff. *Paris 2015: tracking country climate pledges.* 2017. https://www.carbonbrief.org/paris-2015-tracking-country-climate-pledges

Carbon Tracker. *42% of global coal power plants run at a loss, finds world-first study.* November 30, 2018. https://www.carbontracker.org/42-of-global-coal-power-plants-run-at-a-loss-finds-world-first-study/

Carrington, D. Climate emergency: world "may have crossed tipping points." *The Guardian*, November 28, 2019.

ChartsBin. *Atmospheric lifetime of different greenhouse gases*, 2011. http://chartsbin.com/view/2407

Cheng, L., et al. Record-setting ocean warmth continued in 2019. *Advances in Atmospheric Sciences, 37*, 137–142, 2020. https://link.springer.com/article/10.1007/s00376-020-9283-7

Christidis, N., et al. Dramatically increasing chance of extremely hot summers since the 2003 European heatwave. *Nature Climate Change, 5*, 46–50, 2015. https://www.nature.com/articles/nclimate2468

Coady, D., et al. Global fossil fuel subsidies remain large: an update based on country-level estimates. International Monetary Fund working paper WP/19/89, May 2019. https://www.imf.org/en/Publications/WP/Issues/2019/05/02/

Global-Fossil-Fuel-Subsidies-Remain-Large-An-Update-Based-on-Country-Level-Estimates-46509

COP21. *2015 United Nations Climate Change Conference.* 2015. https://en.wikipedia.org/wiki/2015_United_Nations_Climate_Change_Conference

Crowther, T.W. Quantifying global soil carbon losses in response to warming. *Nature, 540,* 104–108, 2016. https://www.researchgate.net/publication/311163076_Quantifying_global_soil_carbon_losses_in_response_to_warming

Deanon, J. *Editor's Vox: methane, climate change, and our uncertain future.* May 11, 2018. https://eos.org/editors-vox/methane-climate-change-and-our-uncertain-future

Dosio, A. Extreme heat waves under 1.5°C and 2°C global warming. *Environmental Research Letters, 13,* 054006, 2018. https://iopscience.iop.org/article/10.1088/1748-9326/aab827

Ehhalt, D., and Prather, M., et al. *Atmospheric chemistry and greenhouse gases. Third assessment report of the Intergovernmental Panel on Climate Change.* 2001. https://www.ipcc.ch/site/assets/uploads/2018/03/TAR-04.pdf

Elder, C.D., et al. Airborne mapping reveals emergent power law of Arctic methane emissions. *Geophysical Research Letters, 47,* doi:10.1029/2019GL085707, 2020. https://agupubs.onlinelibrary.wiley.com/doi/abs/10.1029/2019GL085707

*Emissions for the cement industry.* https://blogs.ei.columbia.edu/2012/05/09/emissions-from-the-cement-industry/

EXXON Research and Engineering Company, Coordination and Planning Division. *$CO_2$ greenhouse effect: a technical review.* April 1, 1982. https://insideclimatenews.org/sites/default/files/documents/1982%20Exxon%20Primer%20on%20CO2%20Greenhouse%20Effect.pdf

Fountain, A.G., et al. The disappearing cryosphere: impacts and ecosystem responses to rapid cryosphere loss. *Bioscience, 62,* 405–415, 2012. https://academic.oup.com/bioscience/article/62/4/405/243859

Frame, C. Another greenhouse gas to watch: nitrous oxide—where are steadily rising levels of the gas coming from? *Oceanus,* December 12, 2008. https://www.whoi.edu/oceanus/feature/another-greenhouse-gas-to-watch--nitrous-oxide/

Free, C.M., et al. Impacts of historical warming on marine fisheries production. *Science, 363,* 979–983, 2019. https://science.sciencemag.org/content/363/6430/979

Fuss, S., et al. Moving toward net-zero emissions requires new alliances for carbon dioxide removal. *One Earth, 3,* 145–149, 2020. https://www.cell.com/one-earth/pdf/S2590-3322(20)30365-1.pdf

Gaudel, A., et al. Tropospheric ozone assessment report: present-day distribution and trends of tropospheric ozone relevant to climate and global atmospheric chemistry model evaluation. *Elementa Science of the Anthropocene, 6*, 1, 39: doi: 10.1525/ elementa.291, 2018. https://www.elementascience.org/articles/10.1525/ elementa.291/

Global Carbon Project. *Supplemental data of Global Carbon Budget 2018 (Version 1.0) [Data set]*. 2018. https://doi.org/10.18160/gcp-2018

*Global Warming Potential Values*. 2014. http://www.ghgprotocol.org/sites/default/ files/ghgp/Global-Warming-Potential-Values%20%28Feb%2016%202016%29_ 1.pdf

Goldenberg, S. Glacier lakes: growing danger zones in the Himalayas. *The Guardian*, 11 October, 2011. https://www.theguardian.com/environment/2011/oct/10/glacier-lakes-melt-himalayas

Goodwin, P.A., et al. Pathways to 1.5°C and 2°C warming based on observational and geological constraints. *Nature Geoscience, 11*, 102–107, 2018. https://www.nature. com/articles/s41561-017-0054-8

Goreau, T.J., and Hayes, R.L. Coral bleaching and ocean "hot spots." *AMBIO, 23*, 176–180, 1994. https://www.researchgate.net/publication/245800342_Coral_ Bleaching_and_Ocean_Hot_Spots

Grace, P., and Barton, L. Meet $N_2O$, the greenhouse gas 300 times worse than $CO_2$. *The Conversation*, December 9, 2014. https://theconversation.com/meet-n2o-the-greenhouse-gas-300-times-worse-than-co2-35204

Grebmeier, J.M., et al. A major ecosystem shift in the northern Bering Sea. *Science, 311*, 1461–1464, 2006. https://science.sciencemag.org/content/311/5766/1461

Grosvold, D.C. *Ozone, the greenhouse gas. 2008.* http://www.aoas.org/article. php?story=20080522125225466

Hansen, J., et al. Assessing "dangerous climate change": required reduction of carbon emissions to protect young people, future generations and nature. *PLoS ONE, 8*, e81648, 2013. https://journals.plos.org/plosone/article?id=10.1371/journal. pone.0081648

Hansen, J., et al. Young people's burden: requirement of negative $CO_2$ emissions. *Earth System Dynamics, 8*, 577–616, 2017. https://www.earth-syst-dynam.net/8/ 577/2017/esd-8-577-2017.pdf

Harvey, F. Melting Antarctic ice will raise sea level by 2.5 metres—even if Paris climate goals are met, study finds. *The Guardian*, September 24, 2020. https://www. theguardian.com/environment/2020/sep/23/melting-antarctic-ice-will-raise-sea-level-by-25-metres-even-if-paris-climate-goals-are-met-study-finds

Hughes, T.P., et al. Spatial and temporal patterns of mass bleaching of corals in the Anthropocene. *Science, 359*, 80–83, 2018. https://science.sciencemag.org/content/ 359/6371/80

Intergovernmental Panel on Climate Change (IPCC). *Fifth Assessment Report, Working Group 1*, 2013. https://www.ipcc.ch/report/ar5/wg1/

Intergovernmental Panel on Climate Change. 2018: Summary for Policymakers. In: Masson-Delmotte, V., et al. (eds.) *Global Warming of 1.5°C. An IPCC Special Report on the impacts of global warming of 1.5°C above pre-industrial levels and related global greenhouse gas emission pathways, in the context of strengthening the global response to the threat of climate change, sustainable development, and efforts to eradicate poverty*, pp. 3–24, Intergovernmental Panel on Climate Change, In Press, 2018. https://www.ipcc.ch/site/assets/uploads/sites/2/2019/05/SR15_SPM_version_report_LR.pdf

International Renewable Energy Agency (IRENA). *Global energy transformation: a roadmap to 2050*. Abu Dhabi, International Renewable Energy Agency, 2018. https://www.irena.org/-/media/Files/IRENA/Agency/Publication/2018/Apr/IRENA_Report_GET_2018.pdf

King, A. Are heatwaves "worsening" and have "hot days" doubled in Australia in the last 50 years? *The Conversation*, June 19, 2017. https://theconversation.com/are-heatwaves-worsening-and-have-hot-days-doubled-in-australia-in-the-last-50-years-79337

*Lazard's levelized cost of energy analysis—version 12.0.* 2018. https://www.lazard.com/media/450784/lazards-levelized-cost-of-energy-version-120-vfinal.pdf

Lelieveld, J., et al. Effects of fossil fuel and total anthropogenic emission removal on public health and climate. *Proceedings of the National Academy of Sciences of the USA, 116*, 7192–7197, 2019. https://www.pnas.org/content/116/15/7192

Lenis Sanin, Y., et al. Adaptive responses to thermal stress in mammals. *Revista de Medicina Veterinaria, 31*, 121–135, 2016. http://www.scielo.org.co/scielo.php?script=sci_arttext&pid=S0122-93542016000100012

Lenton, T., et al. Tipping elements in the Earth's climate system. *Proceedings of the National Academy of Sciences of the USA, 105*, 1786–1793, 2008. https://www.pnas.org/content/105/6/1786

Lenton, T., et al. Climate tipping points—too risky to bet against. *Nature, 575*, 592–595, 2019. https://www.nature.com/magazine-assets/d41586-019-03595-0/d41586-019-03595-0.pdf

Le Quéré, C., et al. Global carbon budget 2018. *Earth System Science Data, 10*, 2141–2194, 2018. https://doi.org/10.5194/essd-10-2141-2018

Li, X., et al. The East Asian summer monsoon variability over the last 145 years inferred from the Shihua Cave record, North China. *Scientific Reports, 7*, 7078, 2017. https://www.nature.com/articles/s41598-017-07251-3

Liu, Z., et al. Global and regional changes in exposure to extreme heat and the relative contributions of climate and population change. *Scientific Reports*, 7, 43909, 2017. https://www.nature.com/articles/srep43909

Louw, A. *Bloomberg new energy finance: clean energy investment trends.* 2017. https://data.bloomberglp.com/bnef/sites/14/2018/01/BNEF-Clean-Energy-Investment-Investment-Trends-2017.pdf

Lovejoy, T.E., and Nobre, C. Amazon tipping point. *Science Advances*, 4, eaat2340, 2018. https://advances.sciencemag.org/content/4/2/eaat2340

Massom, R.A., and Stammerjohn, S.E. Antarctic sea ice change and variability—physical and ecological implications. *Polar Science*, 4, 149–186, 2010. https://www.sciencedirect.com/science/article/pii/S1873965210000411

McNeill, L. This lady scientist defined the greenhouse effect but didn't get the credit, because sexism. *Smithsonian Magazine*, December 5, 2016. https://www.smithsonianmag.com/science-nature/lady-scientist-helped-revolutionize-climate-science-didnt-get-credit-180961291/

Meinshausen, M., et al. The RCP greenhouse gas concentrations and their extensions from 1765 to 2300. *Climate Change*, 109, 213–241, 2011. https://link.springer.com/content/pdf/10.1007%2Fs10584-011-0156-z.pdf

Middelburg, J.J. Marine carbon biogeochemistry. *SpringerBriefs in Earth System Sciences*. https://doi.org/10.1007/978-3-030-10822-9_5

Mora, C., et al. Global risk of deadly heat. *Nature Climate Change*, 7, 501–506, 2017. https://www.nature.com/articles/nclimate3322

Murphy, D.M., and Ravishankara, A.R. Trends and patterns in the contributions to cumulative radiative forcing from different regions of the world. *Proceedings of the National Academy of Sciences of the USA*, 115, 13192–13197, 2018. https://www.pnas.org/content/115/52/13192

NASA Earth Observatory. *Snow cover extent declines in the Arctic.* 2013. https://earthobservatory.nasa.gov/images/80102/snow-cover-extent-declines-in-the-arctic

National Snow and Ice Data Center (NSDIC). *Sea-ice index.* Last accessed April 19, 2021. https://nsidc.org/data/seaice_index/

Oliver, E.C.J., et al. Longer and more frequent marine heatwaves over the past century. *Nature Communications*, 9,1324, 2018. https://www.nature.com/articles/s41467-018-03732-9.pdf

Perkins-Kirkpatrick, S.E., and Gibson, P.B. Changes in regional heatwave characteristics as a function of increasing global temperature. *Scientific Reports*, 7, 12256, 2017. https://www.nature.com/articles/s41598-017-12520-2

Peters, G.P., and Geden, O. Catalysing a political shift from low to negative carbon. *Nature Climate Change*, 7, 619–621, 2017. https://www.nature.com/articles/nclimate3369.pdf

Peterson, T.C., et al. Monitoring and understanding changes in heat waves, cold waves, floods, and droughts in the United States. *Bulletin of the American Meteorological Society*, 94, 821–834, 2013. https://journals.ametsoc.org/doi/pdf/10.1175/BAMS-D-12-00066.1

Post, E., et al. Ecological consequences of sea-ice decline. *Science*, 341, 519–524, 2013. https://www.researchgate.net/publication/254278079_Ecological_Consequences_of_Sea-Ice_Decline

Robine, J.-M., et al. Death toll exceeded 70,000 in Europe during the summer of 2003. *Comptes Rendus Biologies*, 331, 171–178, 2008. https://www.sciencedirect.com/science/article/pii/S1631069107003770?via%3Dihub

Rogelj, J., et al. Paris Agreement climate proposals need a boost to keep warming well below 2°C. *Nature*, 534, 631–639, 2016. https://www.nature.com/articles/nature18307.

Rogelj, J., et al. Mitigation pathways compatible with 1.5°C in the context of sustainable development. In: Masson-Delmotte, V., et al. (eds.) *Global Warming of 1.5°C. An IPCC special report on the impacts of global warming of 1.5°C above pre-industrial levels and related global greenhouse gas emission pathways, in the context of strengthening the global response to the threat of climate change, sustainable development, and efforts to eradicate poverty*, pp. 93–174, Intergovernmental Panel on Climate Change, 2018. https://www.ipcc.ch/site/assets/uploads/sites/2/2019/05/SR15_Chapter2_Low_Res.pdf

Rohling, E.J. *The climate question: natural cycles, human impact, future outlook.* Oxford University Press, Oxford, UK, p. 162, 2019.

Rohling, E.J., et al. Sea surface and high-latitude temperature sensitivity to radiative forcing of climate over several glacial cycles. *Journal of Climate*, 25, 1635–1656, 2012. https://journals.ametsoc.org/doi/full/10.1175/2011JCLI4078.1

Rohling, E.J., et al. Comparing climate sensitivity, past and present. *Annual Reviews of Marine Science*, 10, 261–288, 2018. http://doi.org/10.1146/annurev-marine-121916-063242

ScienceDaily. *Climate change is shifting productivity of fisheries worldwide.* February 28, 2019. https://www.sciencedaily.com/releases/2019/02/190228154846.htm

Shell Greenhouse Effect Working Group. The greenhouse effect. *Report Series HSE 88-001*. Shell Internationale Petroleum Maatschappij B.V., The Hague, The Netherlands, 1988. http://www.climatefiles.com/shell/1988-shell-report-greenhouse/

Sherwood, S., et al. An assessment of Earth's climate sensitivity using multiple lines of evidence. *Reviews of Geophysics, 58,* e2019RG000678, 2020. https://agupubs. onlinelibrary.wiley.com/doi/abs/10.1029/2019RG000678

Smale, D.A., et al. Marine heatwaves threaten global biodiversity and the provision of ecosystem services. *Nature Climate Change, 9,* 306–312, 2019. https://www. researchgate.net/publication/331506733_Marine_heatwaves_threaten_global_ biodiversity_and_the_provision_of_ecosystem_services

Southern California Association of Governments. *Regional transportation plan/ sustainable communities strategy—greenhouse gas emissions and climate change.* 2016. http://scagrtpscs.net/Documents/2016/peir/draft/2016dPEIR_3_8_ GreenhouseGases.pdf

Sun, Y., et al. Rapid increase in the risk of extreme summer heat in Eastern China. *Nature Climate Change, 4,* 1082–1085, 2014. https://www.nature.com/articles/ nclimate2410

The Guardian. *How disappearing sea ice has put Arctic ecosystem under threat.* March 5, 2017. https://www.theguardian.com/world/2017/mar/04/arctic-ecosystem-ice-disappear-ecosystem-polar-bears-fish

Thurber, A.R., et al. Riddles in the cold: Antarctic endemism and microbial succession impact methane cycling in the Southern Ocean. *Proceedings of the Royal Society B, Biological Sciences, 287,* 20201134, 2020. https://royalsocietypublishing. org/doi/10.1098/rspb.2020.1134

Trenberth, K.E., et al. Global warming and changes in drought. *Nature Climate Change, 4,* 17–22, 2014. https://www.researchgate.net/publication/263031929_ Global_warming_and_changes_in_drought

Two Degrees Institute. *Global CH$_4$ levels.* Last accessed date April 19, 2021. https:// www.methanelevels.org

Two Degrees Institute. *Global CO$_2$ levels.* Last accessed date April 19, 2021. https:// www.co2levels.org

Two Degrees Institute. *Global N$_2$O levels.* https://www.n2olevels.org/

UNEP. *Global outlook for ice and snow: highlights.* 2007. https://gridarendal-website-live.s3.amazonaws.com/production/documents/:s_document/232/original/geo-ice-snow-highlights-english.pdf?1487065535

UNEP Global Environment Alert Service. *Where will the water go? Impacts of accelerated glacier melt in the tropical Andes.* 2013. https://europa.eu/capacity4dev/file/ 15905/download?token=yowiV9D7

United States Department of Defense. *Response to congressional inquiry on national security implications of climate-related risks and a changing climate*. 2015. https://archive.defense.gov/pubs/150724-congressional-report-on-national-implications-of-climate-change.pdf?source=govdelivery

United States Energy Information Administration. *How much carbon dioxide is produced when different fuels are burned?* June 17, 2020. https://www.eia.gov/tools/faqs/faq.php?id=73&t=11

Victor,D.G.ProvePariswasmorethanpaperpromises.*Nature,548*,25–27,2017.https://www.nature.com/news/prove-paris-was-more-than-paper-promises-1.22378

Vincent, W.F., et al. Ecological implications of changes in the Arctic cryosphere. *Ambio, 40*, 87–99, 2011. https://www.ncbi.nlm.nih.gov/pmc/articles/PMC3357775/

Wadhams, P. The global impacts of rapidly disappearing Arctic sea ice. *YaleEnvironment360.* https://e360.yale.edu/features/as_arctic_ocean_ice_disappears_global_climate_impacts_intensify_wadhams

Wallmann, K., and Aloisi, G. The global carbon cycle: geological processes. In: Knoll, A.H., Canfield, D.E., and Konhauser, K.O., (eds.). *Fundamentals of Geobiology*, 1st edition, pp. 20–35. Blackwell Publishing Ltd., 2012. http://geosci.uchicago.edu/~kite/doc/Fundamentals_of_Geobiology_Chapter_3.pdf

Welch, L., and Howard, W. *Report of the Defense Science Board task force on trends and implications of climate change on national and international security.* 2011. https://apps.dtic.mil/dtic/tr/fulltext/u2/a552760.pdf

Westbrook, G.K., et al. Escape of methane gas from the seabed along the West Spitsbergen continental margin. *Geophysical Research Letters, 36*, L15608, 2009. https://agupubs.onlinelibrary.wiley.com/doi/full/10.1029/2009GL039191

Wikipedia. *Energy subsidies.* Last accessed May 21, 2019. https://en.wikipedia.org/wiki/Energy_subsidies#cite_note-23

Xu, C., et al. Future of the human climate niche. *Proceedings of the National Academy of Sciences of the USA, 117*, 11350–11355, 2020. https://www.pnas.org/content/117/21/11350

Xu, H., et al. Decreasing Asian summer monsoon intensity after 1860AD in the global warming epoch. *Climate Dynamics, 39*, 2079–2088, 2012. https://link.springer.com/article/10.1007%2Fs00382-012-1378-0

Zanna, L., et al. Global reconstruction of historical ocean heat storage and transport. *Proceedings of the National Academy of Sciences of the USA, 116*, 1126–1131, 2019. https://www.pnas.org/content/116/4/1126

# 3
# The No-brainer
## Emissions Reduction


## 3.1. Setting the Scene

In brutally simple terms, rapid and large-scale emissions reduction is essential. Current human-caused emissions amount to more than 10 GtC (about 40 GtCO$_2$) per year and are still growing. This completely overwhelms the capacity of natural processes for carbon removal into sedimentary rocks. Over the last million years this removal capacity was well below 0.1 GtC per year (Wallmann and Aloisi, 2012). Over the past 200 years, and especially the past 60 years, the anthropogenic carbon addition has even been some 5 to 30 times faster than the most dramatic natural carbon injection we know of since the dinosaurs were wiped out by an asteroid impact. That ancient carbon injection event was due to natural causes, including volcanic activity, and is known as the Paleocene-Eocene Thermal Maximum (PETM; 56 million years ago).

The PETM reached a similar magnitude of carbon injection as business-as-usual scenarios that are maintained over the next two centuries or so. However, during the PETM, it took some 4000–5000 years to reach these magnitudes. And, even more importantly, recovery of the climate system following the PETM took more than 100,000 years. This natural example illustrates the limitations that exist to the timescales over which natural processes alone can clean up such a perturbation. In other words, yes, natural "clean-up" of large emissions will happen, but it will take an exceedingly long time. In consequence, it will fall to us to lend nature a massive helping hand in addressing the problem of humanity's rapidly rising carbon addition to the climate system.

Human "clean-up" assistance needs to come from different angles. First, we all know that it's not much use pumping water from a sinking boat if there is an enormous inflow through a gaping hull breach. Similarly, we cannot hope to restore the climate back to $CO_2$ conditions lower than today if we don't stop the enormous flow of our emissions into the system. This implies a dire need for rapid emissions reduction of both $CO_2$ and other greenhouse gases. Second, even faster-and-faster bailing out of water from our imaginary boat won't help if the hull breach is tearing open further and further at a similar or greater rate. In the same way, activating more and more emissions-reducing measures won't suffice if we allow our total energy hunger and general consumerism, and thus emissions, to keep increasing year on year. In other words, we need strategies for avoiding addition of new emissions associated with the progressive growth and development of society and with humanity's tendency for excessive consumerism. Third, we need to urgently develop and implement methods of carbon removal from the climate system ($CO_2$ drawdown) to help reduce the amount of excess human-caused $CO_2$ that's in the climate system already. This chapter deals with the first and second approaches. Chapter 4 deals with carbon removal.

Section 3.2 discusses $CO_2$ emissions reduction and includes widespread and rapid transition to renewable energy sources, such as wind, solar, and bioenergy, hydrogen when produced without excessive greenhouse gas emissions, hydropower, heat exchange systems, geothermal heat, and so on. One might include nuclear energy if the focus were only on reduction of greenhouse gas emissions, but there is major public aversion to nuclear because of fear for accidents and issues around waste management, security, and potential weaponization. As an all-round solution, the intensively investigated nuclear fusion process appears more interesting because it bypasses the key problem of nuclear waste, but so far it remains a remote dream. Besides the environmental and public perception issues, nuclear plants in all guises are also considered expensive in both installation and operation. As a result, not much growth is generally foreseen in this sector. For example, the 2018 International Renewable Energy Agency (IRENA) report considers a more-or-less stable proportion of nuclear energy in the changing energy mix from now until 2050. But, for energy change required to meet the lower Paris Agreement target of 1.5°C warming by 2100, the same report indicates that major additional measures will be needed to reduce greenhouse gas emissions, although it does not spell these out. There may be an important role for nuclear energy in this; for example, for hydrogen generation in support of industrial and transport processes. In the rest of the book, I avoid the nuclear debate because it is very industry specific, including continued reservations about safety and security for both fission and fusion reactors (Jaczko, 2019).

Section 3.3 explores an array of options for emissions reduction of non-$CO_2$ greenhouse gases, which can contribute considerably to restoring climate change toward safe limits. Methane, $N_2O$, and synthetic greenhouse gases such as CFCs, HFCs, and fluorinated greenhouse gases are of immediate concern. They are influenced by some common factors and by some very gas-specific factors. Because most of these gases have high to very high global warming potentials, reducing their concentrations can make disproportionately large inroads into the necessary adjustment that is needed in the radiative forcing of climate.

Section 3.4 discusses new emissions avoidance, which is often viewed within the remit of emissions reduction. But strategies for emissions avoidance are sufficiently different in nature to deserve separate attention, given that they concern the avoidance of adding new emissions that do not exist yet, rather than reduction of existing emissions. For example, emissions avoidance requires that the annual growth of total energy demand is halted or turned around into a decline. In the current state of affairs, total energy demand keeps growing rapidly each year, and the growth of the renewables sector is not enough to compensate for the combined effect of the total demand increase and—in places—the decommissioning of nuclear power plants. As a result, fossil fuel use fails to decrease. It will be evident that greenhouse gas emissions (including non-$CO_2$ gases) in this sense are part of humanity's wider wastefulness and environmental impact.

Finally, as with all large-scale developments, renewable alternatives have downsides and face challenges to their implementation. These are summarized in section 3.5. Although I will touch upon aspects of policy, governance, and regulation throughout the discussions, I will not discuss the ins and outs of these aspects. This is a pragmatic choice; I do not feel sufficiently qualified to make a real case in these fields and therefore leave them to experts to elaborate.

## 3.2. Reducing $CO_2$ Emissions

In this section, we will discuss key developments with respect to the processes responsible for most of today's human-caused $CO_2$ emissions. These processes range across the power generation and heating+cooling sectors, to the transport, industrial, and agricultural sectors.

Power and heat generation combined are responsible for more than 40% of global carbon dioxide emissions, while transport accounts for about 25%. Agriculture, including land-use change and forestry, is responsible for about 18% of carbon dioxide emissions (its impact is larger, near to 25%, when considering all greenhouse gases). Construction, industrial processes, and waste-related emissions cover the remainder.

Renewable energy has seen profound market penetration in power generation. In other sectors—notably, the heating+cooling and transport sectors—the uptake of renewables appears more challenging and in need of greater emphasis on innovation and on policies to support innovation. Especially in transport, renewable energy covers only a few percent of energy use, with strong dominance of liquid biofuels. Electrification of transport is growing rapidly, but its overall market penetration remains limited as it stands and mostly concerns light vehicles. Heavy transport electrification is in its infancy. Hydrogen is an alternative for replacing fossil fuels or liquid biofuel in the long-haul, heavy-transport sector. Hydrogen also is a prime candidate to replace fossil fuels for heating in the steel industry. Although hydrogen is mostly obtained from fossil fuels, doing so with $CO_2$ capture and storage could make the process carbon neutral. Hydrogen may also be obtained without $CO_2$ emissions through water electrolysis, which is electrical splitting of water molecules, using renewable (or nuclear) energy. Considering heating, only about one tenth of the energy consumed comes from modern renewable energy sources, while about 16% comes from traditional biomass burning.

The field of potential measures of emissions reduction is so diverse that I am sure to miss some remarkable innovations here. Throughout, my synthesis will emphasize those developments that have the potential to deliver particularly large emissions reductions.

## 3.2.1. Power Generation and Heating+Cooling

The Renewable Energy Policy Network for the 21st Century (REN21) Renewables 2018 Global Status Report portrays a global energy transition that is especially well under way in the power generation sector. Renewables

**Table 3.1** Top nations for renewable power.

| Nation | Percentage renewable power (%) |
|---|---|
| Iceland | 100 |
| Paraguay | 100 |
| Albania | 100 |
| D.R. Congo | 100 |
| Namibia | 99 |
| Costa Rica | 98 |
| Tajikistan | 98 |
| Norway | 97 |
| Uruguay | 97 |
| Zambia | 95 |
| Ethiopia | 94 |
| Kenya | 91 |
| Kyrgyzstan | 87 |
| New Zealand | 84 |
| Mozambique | 84 |
| Georgia | 81 |
| Brazil | 80 |
| D.P.R. Korea | 76 |
| Austria | 74 |
| Togo | 73 |
| Angola | 70 |
| Gabon | 68 |
| Panama | 67 |
| Nepal | 66 |
| Canada | 65 |
| Croatia | 65 |
| Colombia | 63 |
| Denmark | 61 |
| El Salvador | 61 |
| Ecuador | 60 |
| China | 25 |
| India | 17 |
| United States | 15 |

Note: Only countries with more than 60% renewable power are shown, along with the "big 3" energy consumers: China, India, and the United States. None of the countries listed has a dominance of biomass burning, which could lead to deforestation. Overall, hydropower dominates.

Source: 2016 values, after Wikipedia, "List of countries by electricity production from renewable sources."

supply a bit more than a quarter of the energy production in this sector. Today, a good few countries operate at high levels of renewable power already, mainly from hydropower, wind, geothermal, and solar (Table 3.1). Among the four top scorers, Iceland produces its power using a combination of hydropower (71%) and geothermal heat (29%), while the other three rely almost entirely on hydropower.

Importantly, things are developing at a rapid pace. Overall, Europe is transitioning most rapidly, with Germany leading the way. Renewables are expected to make up 90% of electricity production in Europe by 2040, dominated by wind and solar (Bloomberg New Energy Finance, 2019).

Investment in new renewable power capacity is now more than double the combined investment in net new fossil fuel and nuclear power capacity (REN21, 2018). As of 2019, modern renewables cover a power capacity of about 2200 Gigawatts, of which about half relates to hydropower. But it's important to emphasize that this is the full potential capacity, which is not always realized. This leads us to discussion of the so-called capacity factor, which is a measure of how much of the total potential capacity is typically achieved. Because the capacity factor is commonly below 100%, an excess of potential capacity must be installed to ensure that enough actual capacity is achieved.

Wind power generation (Figure 3.1a) nowadays reaches a capacity factor of 34% for on-shore facilities and about 45% for off-shore facilities. Solar PhotoVoltaic (PV) systems reach up to 27%, while Concentrating Solar thermal Power (CSP) plants, in which mirrors concentrate the sun's energy to drive engines that create electricity, reach capacity factors up to 45% (Figure 3.1b). Much higher values of 90% or more can be achieved in CSP systems that are coupled to four or more hours of thermal energy storage—usually using molten salt. This is because such systems can draw on their stored energy to keep producing power even when there is no sunlight. Hydropower achieves a capacity factor close to 50% and geothermal power generation around 85%. Bioenergy electricity generation plants—which use combustion of agricultural and forestry residues, biomass gasification, or municipal solid-waste generators—also reach capacity factors up to 85%. For comparison, traditional power generation from nuclear plants, or coal, oil, gas, and biomass burning achieve capacity factors of 70–95%.

Although there will always be impacts of nights and cloudy days and periods of too weak or too strong winds, the last decade has seen substantial

**Figure 3.1** (a): Offshore wind turbines (RGN, 2018) of the Alpha Ventus Wind Farm in the North Sea. (b): Noor Energy 1, a 0.95 Gigawatts solar power project in Dubai, comprising both CSP (circle of mirrors and tower) and PV (rectangular arrays of solar panels).

increases in especially the wind and solar capacity factors. Technological developments will further accelerate this trend, especially through improvement and expansion of energy storage facilities. In fact, technology is no longer limiting the wider implementation of renewables. Delays in new site and capacity development instead arise mainly from administrative issues. Adequate streamlining of administrative review processes will require the adoption of new mechanisms, such as pre-defined renewable energy zones (Luckow et al., 2015).

In recent years, the rate of new solar PV capacity installation was nearly double the rate of wind power capacity installation, and it was also greater than the combined new fossil-fuel and nuclear capacity installation. Installed capacities by 2017 are shown in Table 3.2. It is worth noting that one third of the solar PV capacity was installed in 2017 alone (REN21, 2018), that a further 2 Gigawatts of CSP was under construction, and that only as little as 0.78 Gigawatts of the solar heating capacity was used in direct industrial applications. To put the numbers in Table 3.2 in perspective, global total electricity use was 26,000 Terawatt hours (TWh), which implies a total continuously realized capacity of 2900 Gigawatts.

There is great potential for continued rapid growth in the solar energy sector. And it will be needed; so far, the ever-increasing global energy demand has managed to grow even faster than the increase in total renewable energy supply. In consequence, fossil-fuel use, and $CO_2$ emissions, still increased. In 2018, for example, global energy-related $CO_2$ emissions rose by 1.7%, and the power sector was responsible for almost two-thirds of that emissions growth. It is imperative, therefore, that the growth in renewable energy generation in general—and in solar and wind applications in

Table 3.2 Solar energy capacity by 2017.

| Method | Capacity in 2017 (Gigawatts) |
| --- | --- |
| Solar PV | 402* |
| CSP | 5 |
| Solar heating | 472** |

\* = Direct current (DC)
\*\* = Thermal energy
Source: (REN21, 2018).

particular—be optimized and increased far beyond the current trend if we are to stand any chance of reaching the emission reductions essential for the Paris Agreement climate targets.

Given that solar energy generation is a mature field now, it is ready for dramatic upscaling alongside continued research and development toward further optimization. Renewable power today is the cheapest option for new capacity installation in many parts of the world already, and this cost advantage will increase as prices continue to drop, especially in solar energy generation. Notably, solar and battery prices have plummeted by 85% since 2010. The industry makes economic sense and has the potential to accelerate widespread creation of new jobs over a wide range of skills levels, such as electronics, manufacturing, and engineering. This further strengthens the business case for renewables, with a key role for solar as the heart of the global energy transformation.

Although in a less spectacular manner than solar, wind energy has also been gaining ground at a rapid pace. Since 2010, the costs of wind power generation have dropped 49%. Again, economic drivers are at work as technologies get perfected and implemented at larger scales. The World Wind Energy Association reports that total potential global wind power capacity reached 597 Gigawatts by the end of 2018. Taking into account a capacity factor of about one third, this gives close to 7% of the global electricity demand. There remains sound growth overall, albeit with an apparent reduction in the rate of growth (Table 3.3). Yet, a more interesting and promising picture appears when considering geographic patterns. Although the annual

Table 3.3 Annual growth of global wind power.

| Year | Growth (Gigawatts) |
| --- | --- |
| 2015 | 64 |
| 2016 | 54 |
| 2017 | 60 |
| 2018 | 50 |

Source: World Wind Energy Association. *Wind power capacity worldwide reaches 597 GW, 50.1 GW added in 2018*, February 2019. https://wwindea.org/blog/2019/02/25/ wind-power-capacity-worldwide-reaches-600-gw-539- gw- added-in-2018/

rate of wind-power growth in the traditional markets of Europe reduced over that period, the total capacity still increased. Importantly, though, the rate of expansion has accelerated in Brazil, India, China and other parts of Asia, and some African markets. Notably, China has further cemented its undisputed position as the world's wind-power leader. It has a total wind power capacity of 217 Gigawatts, of which 21 Gigawatts were added in 2018–2019 alone. Far behind China, second place is occupied by the United States with a total capacity of just under 100 Gigawatts, of which 6.7 Gigawatts were added in 2018–2019.

Hydropower relies on the force of moving water. There are four main types of hydropower: run-of-river, storage, pumped-storage, and offshore. Run-of-river hydropower channels flowing water from a river to drive a turbine. It offers little or no storage capacity, but provides a continuous supply of power with some flexibility through regulation of the flow through the turbine. Storage hydropower uses a dam to retain water in a reservoir and generates power by releasing it from the reservoir through a turbine. Storage hydropower offers a lot of flexibility in terms of steady flow and rapid start up and shut down. Pumped-storage hydropower cycles water between a lower and upper reservoir using pumps driven by surplus energy at times of low demand. During high demand, power is produced by releasing water to the lower reservoir through turbines. Offshore hydropower concerns a growing suite of technologies for generating power from tidal currents or wave energy. There are no strict boundaries between the different hydropower systems; combinations and overlaps are common.

As mentioned before, hydropower today provides roughly the same capacity as all other modern renewables combined. The top ten of nations for quantities of hydropower production include China, Brazil, Canada, United States, Russia, Norway, India, Venezuela, Sweden, and Japan. China alone produces more hydropower than Brazil and Canada combined. Hydropower makes up by far the largest proportion (96%) of Norway's energy production. Though Norway (Figure 3.2) has become a global posterchild for hydropower, it's worth noting that Paraguay, Albania, the D.R. Congo, and Tajikistan have higher renewables percentages, and that these are all dominated to larger extents by hydropower than in the case of Norway (Table 3.1).

Tidal power generation employs the energy of tides by driving turbines with tidal inflow and outflow currents through ducts in a tidal barrage (dam). Although these are hydropower types with great future potential,

**Figure 3.2** The 100 million Watts Tyssedal hydropower station and museum in Norway, sympathetically integrated within the Sørfjorden landscape. This is one of Norway's almost 1000 hydropower stations.

*Source*: David Aasen Sandved, CC BY-SA 3.0, Wikimedia Commons. https://en.wikipedia. org/wiki/Tyssedal_Hydroelectric_Power_Station

preciously few major commercial facilities exist yet, though many small systems are operational. Among the large commercial facilities, the Rance Tidal Power Station in the Rance River of Brittany, France, for a long time reigned supreme as the tidal hydropower system with the largest installed capacity in the world (Figure 3.3). Opened in 1966, it was the world's first tidal power station, with a total installed capacity of 240 Megawatts, albeit with a capacity factor of only 28%. In 2011, it was surpassed in installed capacity by the 254 Megawatts South Korean Sihwa Lake Tidal Power Station, although this only achieves an even lower capacity factor of 25%. In the United Kingdom, the Severn estuary has a very large tidal range, and a potential tidal barrage in that estuary has been under discussion for decades. Tidal stream generators are an alternative to tidal barrages. These are free-standing, or floating, units that function like underwater wind turbines. However, most only have capacities of 1 Megawatt or less.

**Figure 3.3** Rance Tidal Power Station in the Rance River of Brittany, France.
*Source*: Tswgb, Wikipedia. https://en.wikipedia.org/wiki/Rance_Tidal_Power_Station#/
media/File:Barrage_de_la_Rance.jpg

Other hydropower technologies that need to be mentioned remain experimental only, but are thought to have major potential. For example, wave energy farms can be either offshore or nearshore. Offshore ones are generally considered most promising for high energy production, but current prospects for commercialization seem limited. Together with many other marine energy recovery systems, the different techniques and technological readiness stages for wave energy systems are summarized extensively in the Marine and Hydrokinetic Technology Database. This database also contains details on ocean thermal exchange conversion (OTEC) systems. These divide into two broad categories. Closed OTEC systems utilize the temperature gradient between surface and deeper waters in the oceans through a complex of heat exchangers and a working fluid, which is used to drive a turbine for electricity generation. One of the heat exchangers is for cooling the working fluid and relies on heat exchange with pumped-up cool deep waters. Open OTEC systems pump warm sea-water directly into a low-pressure container, where it boils because of the low pressure. The vapor is then used to drive a turbine and condensed—producing drinkable freshwater—in a heat exchanger that is cooled with cool deep water. OTEC plants can be based on shore but are more efficiently located offshore, either on legs in shallow water or floating in the deep ocean. OTEC plants have typical capacity factors of 80–100%.

OTEC plants are most effectively placed within limits of about 20° of latitude from the equator because those tropical regions combine the presence of abundant warm surface water (25–30°C) with relatively easy access to colder deeper waters (5–10°C). This easy access to colder deeper waters arises from equatorial-zone upwelling of deeper waters toward the surface, driven by the large-scale structure of ocean circulation. Even in the tropical regions, however, the temperature contrast between surface and deep waters remains relatively modest. In consequence, OTEC plants require very large volumes of warm surface water and cool deep water to extract sufficient heat for generating constant renewable power. Makai Ocean Engineering has an operational 100 kilowatts OTEC test rig in Hawaii (Figure 3.4) and is considering future offshore rigs that will be good for a capacity of 200 Megawatts. There is a wealth of other proposed projects globally, all of which fall in the 1–100 Megawatts range. The capacity for drinking-water provision by these systems is often as important as their capacity for energy provision.

Geothermal energy is obtained from the heat gradient in Earth. Temperatures are very high deep inside Earth, mainly as a result of the ongoing decay of radioactive elements. The heat of that process escapes very gradually toward shallower layers of the Earth and drives tectonic movements, volcanic processes, geysers, heat vents, and so on. In other words, geothermal heat is an expression of the gradual cooling of Earth from the inside out. In some places, notably regions with higher volcanic activity, the heat comes naturally to the surface and is more easily accessible. In other places, the heat can be accessed by drilling down a few kilometers. The use of geothermal energy comes in two guises: geothermal power generation uses the heat in different ways to drive turbines that generate electricity, while geothermal heating uses the heat directly.

Geothermal power generation (Figure 3.5) has a global average capacity factor of about 75%, though a value as high as 96% has been reported in the United States. The high capacity factor of geothermal power reflects the constancy of the geothermal heat flux. Its lack of intermittency sets geothermal power generation apart from wind and solar and makes geothermal power highly suitable for base-load power provision. As of 2018, about 14 Gigawatts of geothermal power capacity was installed globally. The global potential for

**Figure 3.4** Makai Ocean Engineering's first US-grid-connected, operational, 100 kilowatts OTEC rig in Hawaii.

*Source*: Vitafougue, CC BY-4.0, Wikimedia Commons. https://commons.wikimedia.org/wiki/File:Makai%27s_Ocean_Thermal_Energy_Conversion_Scale_Up.jpg

geothermal power generation is estimated at about 200 Gigawatts. Reaching this would require continued technological advances and new approaches, and the estimated maximum potential also depends strongly on cost and incentive developments.

**Figure 3.5** The Nesjavellir geothermal power plant in Iceland.
*Source*: Gretar Ívarsson, Wikimedia Commons. https://upload.wikimedia.org/wikipedia/commons/9/9f/NesjavellirPowerPlant_edit2.jpg

For geothermal heating, capacity factors of 13–73% are reported, with a global average value of about 30%. Although the heat is "always on," it is still reported with a relatively low average capacity factor because the demand for heat is intermittent. Most geothermal heat is used for bathing (45%) and direct heating of spaces (34%), though agriculture also is an important user in some countries—primarily for heating greenhouses. Geothermal heat pumps, or ground-source heat pumps, are a special category within geothermal heating, in that they don't really use geothermal heat. Instead, they rely on the fact that below 6 meters, the undisturbed ground temperature is consistently at the mean annual air temperature. Many people know this from the temperatures felt when visiting deep caves. In consequence, the undisturbed ground is cooler than the air in summer and warmer than the air in winter, so that heat exchange can be used for cooling in summer and heating in winter.

Most installed geothermal energy capacity currently exists in the United States, though Iceland rules the league table for the geothermal proportion within its total national energy production. But it is in The Netherlands that we find Europe's most rapidly expanding market for geothermal energy use. This is the result of specific government policy instruments to facilitate such a rapid development (Willemsen, 2016). First, the creation of the openly accessible Netherlands Oil and Gas subsurface data eliminated the need for exploration drilling. Second, the government set up a fund to mitigate for geological risks within geothermal projects. Third, a feed-in premium was set for renewable heat, to stimulate transition from fossil-fuel energy to renewable energy. Such programs of facilitation and incentivization create a healthy environment for growth in this exciting field of versatile, non-intermittent, renewable energy provision.

Bioenergy electricity generation plants use combustion of biomass. Biomass is biological, organic material from living or recently deceased organisms. The idea is that combustion of biofuels would be emissions neutral in that their growth consumed as much $CO_2$ as is released during their combustion. Globally, bioenergy accounts for about 95% of renewable energy supplied to direct heating and about 100 Gigawatts of the global total of 2200 Gigawatts of potential renewable power generation capacity. Average capacity factors for bioenergy power production range from 63% in China to 83% in North America. Higher capacity factors are typically found for larger projects. Biofuels are also potentially useful emissions-neutral replacements for fossil fuels in transport, with applications ranging from trucks and cars to planes.

Biofuels (Figure 3.6) can be solid, liquid, or gaseous. Important bioenergy sources are agricultural and forestry residues—either solid or in liquefied form such as bioethanol and biomethanol—gasified biomass, and municipal solid waste. Wood remains the most common source of bioenergy. Its use ranges from traditional burning in log fires to modern wood-pellet combustion systems. Liquid biofuels experienced a surge in popularity in the late 1990s to early 2000s, when bioethanol and biodiesel from oil palms, sugarcane, eucalypts, and other crops, as well as discarded vegetable oils and animal fat, were advanced as renewable or so-called green alternatives to fossil fuels. Gaseous biofuel is dominated by methane that is released upon decomposition of organic matter, which can range from clean biomass to household waste. Pushing biomass energy production to the limits—while accounting for realistic technological progress, realistic use of residues from forestry and agriculture and organic waste, and avoidance of competition with food production for prime land—it might be possible to achieve a total

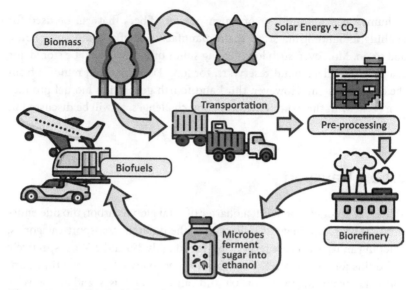

**Figure 3.6** The biofuel life cycle.

*Source*: Following Qidwai, A., et al. Introduction of nanotechnology in the field of biofuel production. In: Srivastava, N., et al. (eds.). *Green nanotechnology for biofuel production, biofuel and biorefinery technologies 5*, pp. 29–38, Springer International Publishing AG, Switzerland, 2018. https://www.researchgate.net/publication/323849526_Introduction_of_Nanotechnology_in_the_Field_of_Biofuel_Production

annual bioenergy potential of 40–100% of the current annual global energy demand (Faaij, 2012).

So far, two basic generations of liquid or gaseous biofuel development can be distinguished (Davidson, 2008). The first generation concerned mainly ethanol from dedicated plant crops. The second generation derived from woody, inedible parts of plants, such as grasses, crops, and forest waste, and also animal waste. A third generation of biofuels, which is currently being evaluated, is based on liquid or gaseous biofuels from cultured algae. This would help to avoid the detrimental environmental impacts that are associated with widespread implementation of the earlier generations. And finally, there is the so-called fourth generation of biofuels, production of which relies on genetically engineered microbes for optimized photosynthesis. Incidentally, this fourth-generation approach has much broader potential than just biofuel generation, including also the generation of food for animal or human consumption, or of tough polymer fibers that can be used for clothing, as building materials, in the manufacturing of cars and even planes, and so on. Moreover, such long-lasting solids or liquids could be stored, for example in underground reservoirs, for long-lasting carbon removal from the climate system. However, third and fourth generation biofuel production come with their own environmental challenges, as will be discussed in section 3.5.

## 3.2.2. Transport

Transport accounts for about a quarter of total global carbon dioxide emissions. This breaks down as follows over the different transport categories. Shipping and aviation are responsible for roughly 3% and 2.5%, respectively and trains for less than 1%. The other 19% or so arise from road transport, with cars accounting for about 10% and vans, trucks, buses, and motorcycles together making up about 9%.

Many fuel/energy changes being considered for emissions reduction in transport span several or all of the transport categories. With respect to road transport, almost everyone will be familiar to some extent with electric and hydrogen fuel-cell technologies. In future, these may also be used in a hybrid form, where a hydrogen fuel cell is used to charge batteries, similar to how a combustion engine charges batteries in many hybrid vehicles today. Nearly everybody will also be aware of biofuel use in road transport, either in the form of admixtures to fossil fuels (e.g., E10) or in undiluted form (e.g., biodiesel). Biofuels may have a great role to play in transport, given that their application alone could potentially reduce greenhouse gas emissions to about a tenth of what they would be from conventional fossil fuels (Muller-Langer et al., 2012). But to realize this potential, new means of biofuel generation will be needed that overcome the drawbacks of first- and second-generation biofuels.

On the railways, all of the highly traveled networks are electrified. Therefore, renewable sources of electricity constitute the majority of alternative fuels in rail transport. In the European Union, only 20% of railway operations comes from diesel traction, and this could be replaced by biodiesel. In the specific case of The Netherlands, all electric trains of the national railway company Nederlandse Spoorwegen, which together transport some 600,000 people per year, are claimed to have been running on energy harvested from the wind since January 1, 2017. Over a year, the energy involved is about equivalent to that needed to power the city of Amsterdam. In a similar manner, solar power could replace fossil-fuel derived power in electric rail networks elsewhere. In addition, direct solar powering of trains, where each train provides its own power, may not be a major development, but still can be done under the right circumstances. The Byron Bay Railroad Company in Australia brought a solar-PV powered train into service in late 2017. It consists of an upgrade of the original two 1949 train carriages from diesel locomotion to battery-electric locomotion, powered predominantly by PV panels carried on the roof. It carries up to 100 passengers over a 3-kilometer return route on a 1-hour schedule.

Beyond electric railway operations, diesel reigns. The global consumption of liquid fuel by railway transport is of the order of 34,000 million liters. This would suggest that biodiesel will be a critically important fuel for the future, but the use of biodiesel in trials has revealed substantial issues, such as increased $NO_x$ emissions (Skinner et al., 2007). This is true for both for railway use and heavy road use of biodiesel; soybean-based biodiesel appears to be worse in this respect than rapeseed- and animal-based biodiesel. Moreover, there are significant $N_2O$ emissions in the course of the production of the biodiesel feedstocks, as well as a serious issue of competition with food production for land-surface and water and nutrient resources. Finally, operational issues have been reported when running normal diesel engines in trials with fuel blends that contain more than 20% biodiesel mixed with conventional diesel.

Hydrogen is also considered an important future renewable fuel for railway transport. In contrast to a considerable number of hydrogen-powered light and heavy road vehicles, hydrogen-powered trains remain exceedingly rare. In September 2018, the first two hydrogen-powered trains to enter commercial service were launched in Germany. Built by the French manufacturer Alstom, they service a 100-kilometer route. Hydrogen power requires specific infrastructure. It also requires specific methods for production of the hydrogen in order to count as a zero-emissions fuel. I address these issues later in this section.

Shipping requires fuels that can be stored in large quantities and that have enough energy density to sustain the heavy transport over long distances typical of this sector. Conventionally, shipping has relied on heavy fuel oil and marine diesel oil. An existing drive toward reduction of sulfur, nitrous-oxide, and particulate-matter emissions has caused increasing uptake of liquefied natural gas (methane) instead of heavy fuel oil. However, liquefied natural gas only offers up to 20% $CO_2$ emissions reduction relative to conventional fossil fuels. It also requires considerable shore- and ship-based infrastructure and modifications, along with three to four times greater fuel storage capacity than diesel. Finally, leakage of natural gas needs to be carefully controlled because methane is a powerful greenhouse gas. One particular problem in this respect is methane slip, which is the release of a fraction of unburned methane from the engines.

Other fuels considered for shipping include methanol, ethanol, biofuels (including liquefied biogas), liquefied hydrogen and ammonia, and synthetic diesel-like fuels. The latter are human-made fuels from reactions that combine hydrogen with $CO_2$ or carbon monoxide, with potential uses in both shipping and other means of transport. Synthetic fuel generation can be done in reactors that require energy, or—as a matter of specific interest—through artificial photosynthesis that uses only light as the source of energy, imitating the processes operated by plants. Scientists have been able to make artificial photosynthesis work reliably, but its efficiency is still nowhere near that achieved by plants.

Electrification is increasingly considered for ships, but current limitations to power storage capacity still make it a feasible option only for shorter-distance applications such as ferries, with Norway leading the way. In addition, hybrid drives of various types are being developed and implemented, combining liquid or gaseous fuels with electric drives and battery storage, in conceptually similar ways to hybrid cars. Finally, wind and solar power are being assessed in shipping, but mostly for additional power rather than as a main power source. Combinations of wind and solar power have also been proposed, with solar PV panels embedded in metal sails. Across all potential energy sources, it is important that emissions are examined across the entire life cycle. That means the entire cycle from recovery of the fuel to emissions to drive a car's wheels or a ship's propellers. This is referred to as "well to drive/propeller," which is often separated into "well-to-tank" and "tank-to-drive/propeller" emission values.

In aviation, fuel efficiency increases of 30–35% are thought to be possible from modern gas turbo engines, which will help reduce emissions. But much more is needed for aviation to go down to zero emissions; this underlies a major push toward alternative jet fuel development. Alternative fuels will have to meet current specifications for jet fuel, either on their own or in blends with conventional jet fuel. In consequence, the sector is most focused on development of suitable biofuels. In 2018, about 15 million liters of aviation biofuel was produced, but this accounted for not even 0.01% of the total aviation fuel demand.

Electric propulsion is also an area of considerable interest in aviation and spans three main categories: all-electric, hybrid electric, and turboelectric. All-electric systems use only power from batteries on the aircraft. Hybrid systems use gas turbine engines both for propulsion and to charge batteries, while the batteries can provide propulsion energy during certain flight phases. In turboelectric systems, there are no batteries to provide propulsion energy during any phase of flight. Instead, these systems use gas turbines to drive electric generators, which power individual electric fans. Because of today's battery limitations, only turboelectric systems are currently earmarked for high-priority development in the electric propulsion sector, to further curtail $CO_2$ emissions from aviation over the next 10–30 years.

* * *

Used throughout the transport sector, ethanol and methanol fuels include bioethanol and biomethanol, which are produced from biological waste products and thus are more or less $CO_2$ neutral when combusted. Similar to liquefied natural gas and hydrogen, ethanol and methanol have limited energy density; less than half that of conventional fossil fuels, so that about 2.5 times more storage volume is needed. Methanol has advantages in that it is a liquid that can be transported in tankers at atmospheric temperature and pressure, and stored in tanks similar to those used for gasoline. In consequence, the distribution, handling, and bunkering systems for methanol would be very similar to those used today for conventional fuels. Methanol also poses a hazard, however, in that its fumes are highly poisonous and, if inhaled, can lead to blindness and death.

Hydrogen is seen as a critically important fuel for the future; its combustion produces only water vapor (Figure 3.7). Today, hydrogen is mainly produced from fossil fuels. The most common method is steam reforming of natural gas, which produces $CO_2$ along with the hydrogen. Other fossil-fuel methods produce carbon monoxide or activated carbon along with the hydrogen. Water electrolysis is not a major source of hydrogen yet, but it may

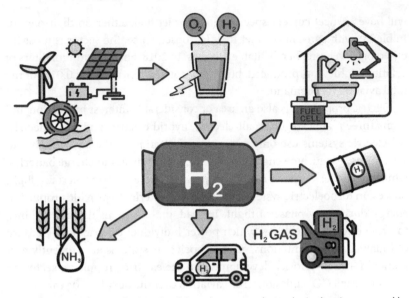

**Figure 3.7** Hydrogen fuel cycle, driven by water electrolysis that is powered by renewable energy. The only emission is water vapor.

*Source*: Based on Ariizumi (2010). https://ourworld.unu.edu/en/renewable-hydrogen-key-to-a-new-civilization

become so in the future, and can be free from $CO_2$ emissions if powered by renewable or nuclear energy. Finally, pilot-study sites exist for hydrogen production via thermochemical conversion of biomass; this does produce $CO_2$ emissions, but these would be more or less offset by $CO_2$ uptake during biomass growth.

Hydrogen has the great advantage of being a fuel that can be transported over long distances by established means, such as pipelines, ships, and tankers. It can also be stored underground or in containers. Liquefied hydrogen produced from natural gas *without* $CO_2$ capture and storage has high well-to-tank emissions, but zero tank-to-drive/propeller emissions. Production of liquefied hydrogen *with* $CO_2$ capture and storage, or via water electrolysis using renewable energy, would eliminate the well-to-tank emissions. In that case, hydrogen becomes a true zero-emissions fuel.

For transport, Japan has a special interest in transitioning to hydrogen, based on its limited space for large-scale renewable energy development. As it stands,

Japan's vision for this transition hinges on hydrogen production from brown coal in Australia, with $CO_2$ capture and storage into underground reservoirs to make the process $CO_2$ neutral (aside from the mining). Hydrogen can also be used for power generation; for example, a natural gas turbine is being modified for pure hydrogen combustion by 2024 at the Vattenfall's Magnum power plant in Groningen, The Netherlands. This will again be fed from fossil fuels, given that the hydrogen will be derived from Norwegian natural gas, with $CO_2$ capture and storage into natural caverns. In both initiatives, access to economical $CO_2$ capture and storage techniques is the limiting factor. South Korea is another country with profound interest in a hydrogen economy, in this case including transport as well as household and commercial power generation.

On March 15, 2018, the future of hydrogen in the next-generation global energy mix got a considerable boost. On that day, China's State Council announced a push toward the development of fueling stations for hydrogen fuel-cell cars, in a bid to diversify its carbon-free transportation. This decision is important because of a revealing precedent: China previously managed to use subsidies and other policy incentives to create—within a decade—the world's largest market for battery-powered electric vehicles. Hydrogen generation in ways that avoid $CO_2$ emissions will be essential for all these initiatives, either through $CO_2$ capture and storage or through renewables-based approaches.

Ammonia is also touted as a particularly promising fuel for the future. Each ammonia molecule ($NH_3$) contains more hydrogen atoms than a hydrogen molecule ($H_2$). It can be burned either in mixes with other fuels in a variety of engines, or by itself in adapted conventional gas turbines. When ammonia is burned as a fuel, it returns to nitrogen and water, both of which are harmless and very common in nature. Ammonia also has its own leak-detection property: it is easily detected by smell. Finally, a high-volume global ammonia manufacturing and distribution infrastructure exists already for the fertilizer industry. In fact, ammonia production is one of the world's biggest chemical industries, and ships move it around the world on a daily basis. Those ammonia transporting ships are perfect early adopters of ammonia as a fuel, by using some of their cargo. Beyond shipping, ammonia is also considered for energy storage linked to concentrated solar power plants. In other words, ammonia can not only be made using any renewable source of electricity, but it can also be used to store and transport any energy obtained from renewables.

Hydrogen generation either via ammonia, or directly, during periods of surplus renewable energy supply could provide sustainable fuel to provide base-load energy security through periods when insufficient renewable energy is available. Moreover, it is not only a potentially important fuel source for long-haul, heavy transport and light vehicles, but potentially also for longer-range commercial jets by around 2050.

## 3.2.3.  Steel and Cement

In the steel industry, hydrogen is a prime candidate for replacing fossil fuels. $CO_2$ emissions from present-day steel-making processes based on coal or charcoal reach 1.7–4 tons of $CO_2$ per ton of steel, with a global average of 1.83 tons of $CO_2$ per ton of steel in 2017. The industry thus generated roughly 8% of the global anthropogenic $CO_2$ emissions from fossil-fuel combustion. To begin with, important gains may be made by introducing stricter circular economies for steel, with more recycling. But new technologies, with a key role for hydrogen, will also be needed to push emissions toward zero. In Sweden, the Hydrogen Breakthrough Ironmaking Technology aims to produce zero-carbon steel from 2020 onward. This is not an isolated initiative; many other steel makers are investigating the use of hydrogen as well, and work to their own deadlines for zero-carbon steel production. For example, ThyssenKrupp aims to achieve zero-carbon steel production by 2050. Another suggested new approach to steel production, the iron ore electrolysis or Electrowinning (EW) approach, uses electricity directly as reducing agent. This method is still in its infancy and market entry is not expected before 2040. In hydrogen-based steel production approaches, (renewable) energy is needed to produce the hydrogen. In the EW approach, electricity is used directly, without the intermediate step, and it therefore is the most energy efficient approach (Fischedick et al, 2014).

Concrete is the most widely used man-made material and is the second-most-consumed resource globally after water. Cement production is responsible for about 8% of total global $CO_2$ emissions, arising both from the transformation of limestone into clinker and from the energy generation needed for that process. Some initiatives target net $CO_2$ emissions reductions within the framework of conventional cement production. These focus on the use of alternative fuels or $CO_2$ capture and storage. Alternative energy production could be achieved from biofuels, or biochar from mobile fast-pyrolysis units; biochar is a form of charcoal that is produced by heating biomass from sustainable sources to very high temperatures in the absence of oxygen (pyrolysis). But it is critical that such energy transitions for cement production are preceded by impact assessments to determine if the proposed fuel pathways are economically feasible. Also, these assessments need to establish that sustainable feedstocks are available in sufficient quantities without major detrimental effects on food production or biodiversity. Further efficiency may be achieved by biochar admixture into cement mortar, which looks promising for improving cement characteristics, waste recycling (biochar from waste saw dust), and reduction of cement demand.

Other initiatives instead focus on the development of alternative, novel cements and masonry. Many of these are inspired by geological or biological cement-forming processes, and some concentrate on recycled waste incorporation. Production of these novel materials ranges from zero-emissions approaches to methods that even capture/extract additional $CO_2$ from the air, sequestering it long term in the building materials. Another consideration is that cement/concrete is not always needed, and a transition from conventional, energy-intensive, to alternative building systems in many applications might more-or-less halve the embodied energy of buildings.

## 3.2.4. Agriculture

Agriculture, including forestry, today contributes about a quarter of all greenhouse gas emissions (Figure 3.8). Not all of these emissions are $CO_2$; there are intense emissions of methane and $N_2O$ as well. More extensive discussion of non-$CO_2$ greenhouse gases will follow in section 3.3.

Key to reducing the sector's climate-change impact is making agriculture less industrial/intensive, or less reliant on heavy greenhouse-gas-producing processes and chemicals such as pesticides and fertilizers. To achieve this, some initiatives consider a return to multi-cropping and other more traditional, regenerative approaches and practices that stimulate natural nutrient cycles and resistances (for example, Kittredge, 2015; Regeneration International, 2019). Regenerative agriculture reduces the burden of dependence on chemicals. For example, synthetic fertilizers are replaced by diverse crop rotations, no-till planting, and management of livestock grazing.

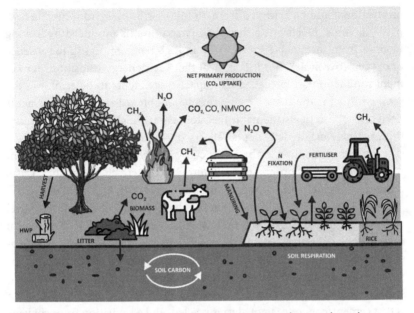

**Figure 3.8**  The main greenhouse gas emission sources/removals and processes in agriculture. CO stands for carbon monoxide, HWP for harvested wood products, and NMVOC for non-methane volatile organic compounds.

*Source*: Based on Paustian et al. (2006). https://www.ipcc-nggip.iges.or.jp/public/2006gl/pdf/4_Volume4/V4_01_Ch1_Introduction.pdf

Reported outcomes of regenerative agriculture include re-enrichment of soils with carbon (which, incidentally, locks carbon away from the climate system), improved soil moisture retention, reversal of desertification and soil erosion, increased biodiversity, reduced greenhouse gas emissions, and production of nutrient-rich food. It may also prove essential for building resilience against increasing stress on crops and livestock from climate change impacts, because diversity is a safer bet for continuation of some essential food production than large-scale monocultures that are sensitive to single pests or stress tolerances.

On the other end of the spectrum, the search for less industrial/intensive agricultural practices has drawn attention to novel methods, including the potential use of biomass formed with certain (autotrophic aerobic) $CO_2$-binding microbes that are fed with $CO_2$ from industrial sources (Figure 3.9). The same can be done with methane-utilizing bacteria or with yeasts grown on wood or

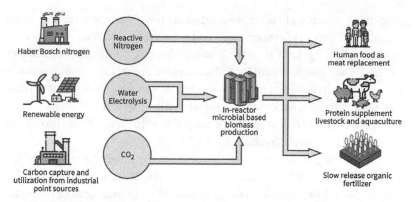

**Figure 3.9** Microbial biomass production and utilization. Carbon emission avoidance and capture by in-reactor production of microbial biomass based food, feed, and slow release fertilizer.

*Source*: Based on Pikaar et al., Potentials and limitations, *Science of the Total Environment*, *644*, 1525–1530, 2018. https://www.sciencedirect.com/science/article/pii/S004896971832 5749?via%3Dihub

seaweed. The produced microbial biomass is commonly known as microbial protein because it has a crude protein content of 70–75%. Microbial protein can serve as a protein supplement in animal diets, including fish food for aquaculture, or as a slow-release organic fertilizer that provides both organic nitrogen and carbon to agricultural soils. Microbial protein meets the regulatory requirements for human nutrition and therefore presents an avenue toward meat replacement for human consumption.

While this might sound at first like science-fiction space-food technology, most of us will recognize at least some microbial protein applications in food production. For examples, consider the use of fungi, yeast, bacteria, and algae in production of foods such as bread, yoghurt, mushrooms, and beer. Furthermore, microbial fertilizer development provides three really interesting additional benefits. First, it improves soil fertility and thus enhances crop yields. Second, it can reduce the widespread use of nitrogen-based fertilizers, which drive major releases of the powerful greenhouse gas $N_2O$. Third, organic matter in soils helps moisture retention, which is particularly important in marginal, arid regions.

Overall, in conventional agriculture, emissions reduction is easiest to achieve in land-rich countries with low population densities and extensive agriculture, which includes major emissions from deforestation and forest degradation. Large reductions are possible in these emissions without substantially affecting food security because there is significant potential in land-rich countries to intensify production on existing crop- and grasslands in greenhouse gas efficient manners. Conversely, in densely populated countries with extensive agriculture, a forceful drive toward emissions reduction bears a serious risk of causing a steep rise in the number of chronically undernourished people (Frank et al., 2017).

As mentioned at the outset of this section, the suite of highlighted examples for emissions reduction is not exhaustive. There are many more developments in low-emissions energy generation and industrial or domestic processes, or $CO_2$ emissions-reduction or capture from other processes. For example, important improvements are afoot in insulation of domestic and workspaces with products that have very low emissions footprints, or that even represent net $CO_2$ removal from the climate system. In selecting these examples, I have merely focused on those trends in the key energy, transport, industrial, domestic, and food-production branches that show particular potential for driving large-scale reductions in $CO_2$ emissions.

## 3.3. Reducing Non-$CO_2$ Greenhouse Gas Emissions

$CO_2$ is not the only greenhouse gas; methane, $N_2O$, and the synthetic gases such as CFCs, HFCs, and fluorinated greenhouse gases are also of concern. These gases have higher global warming potentials than $CO_2$, so that emissions reductions, emissions avoidance, or removal of a certain quantity of these gases is even more effective at combating global climate change than such actions for the same quantity of $CO_2$. As such, these other gases are sometimes referred to as the low-hanging fruit, or the easy pickings. While this sort of term may have a nice ring to it, true practical application is another matter altogether.

A great deal of non-$CO_2$ greenhouse gas management is closely related to $CO_2$ management and to land-use changes that by themselves also have climate change impacts. A quarter of all greenhouse gas emissions comes from agriculture (together with forestry), producing food for humans either directly or indirectly through livestock. For example, livestock rearing leads directly to methane and $N_2O$ releases, but also impacts on climate change through deforestation for expanding pastures. The deforestation releases carbon that was locked away in the forests, in the form of either methane or $CO_2$. It also leads to soil degradation, with release of soil carbon into the climate system, again as methane or $CO_2$. But agriculture is by no means the only source, even though it's a major one. In this section, I therefore discuss key aspects of non-$CO_2$ greenhouse gas emissions in a broader sense, along with methods and new concepts for their reduction. I start with methane, for which I draw heavily—but not exclusively—on the recent comprehensive review by Nisbet et al. (2020). $N_2O$, $NO_x$, and the synthetic greenhouse gases then follow.

\* \* \*

While $CO_2$ accounts for about half of the human-caused greenhouse-gas-related forcing of climate change since the onset of the industrial revolution, methane accounts for about a quarter. In general, methane emissions come from two different types of sources. One type comprises diffuse sources that are very difficult to pinpoint and not at all easy to capture and the other consists of discrete sources that are—in principle—more easily located and captured. Diffuse sources include, for example, methane release from melting permafrost or from forest or grassland fires. Discrete sources include leaks from natural gas infrastructure, wastewater and sewage, landfills, and shafts and ventilation outlets of underground coal mines.

Emissions from discrete sources may be reduced in relatively straightforward ways, often with economic or safety benefits. Methane leaks from natural gas infrastructure span a wide range across inadvertent leaks, deliberate flaring, venting, maintenance, so-called slippage of unburned fuel during combustion, and so on. The most effective reduction of emissions from natural gas infrastructure requires prioritizing the termination of the largest leakages from wells, pumping stations, and pipelines and reducing flaring and venting. The next level of improvements requires repairs to the worst leaks in urban distribution networks, supported by longer-term maintenance and replacement programs. Because methane leaks are very dangerous, a prioritized program of repairs will greatly improve safety as well. Slippage of unburned fuel during combustion can be reduced by catalytic methane removal. A recently developed catalytic boiler manages to achieve 0 ppm methane emissions, which is even lower than atmospheric methane concentrations.

In wastewater treatment plants and agricultural settings, methane can be recovered at efficiencies of 50–100%, using different types of bioreactors for methane oxidation. Concerning solid waste–based emissions, these are important everywhere from landfills, especially in less developed or regulated countries. Most of the gas escapes from the working face of a landfill, which is necessarily uncovered. One solution is to rapidly cover landfill working faces with soil that hosts methane-oxidizing bacteria. This oxidation potential can be further increased by enriching the landfill soil cover with biochar.

In underground coal mining, methane escapes both during mining and surface processing and during methane extraction from coal beds. Gas-rich air is infamous for causing explosions and venting is essential, often into combustion chambers. Combustion turns the methane into $CO_2$, which molecule for molecule has a much lower greenhouse warming potential than methane. Thorough assessments are needed per case to evaluate the global warming impact through the entire chain of mining, venting, and combustion, but releasing the gas as $CO_2$ likely causes less global warming impact than releasing it as methane. But let's be very clear about one thing: methane combustion could only reduce the total greenhouse warming potential of emissions to some extent; it will not at all suffice for bringing the global warming potential anywhere near to zero. The latter could be achieved only if the methane (or $CO_2$) were captured and stored away from the climate system.

Several novel concepts and methods have also emerged for dealing with methane emissions from local point sources, such as landfill, wastewater, and mine sites, or industrial leakages. Some have made it into implementation already. For example, some methods focus on the use of bacteria known as methanotrophs, which digest methane and convert it into products such as formaldehyde, which in turn can be used to manufacture hydrogen fuel. Efforts in genetic engineering are afoot that attempt to improve the efficiency of methanotrophic bacteria. Alternatively, some projects aim to develop artificial ways of mimicking methanotrophic bacteria, with a view to scaling up such artificial applications for direct removal of large quantities of methane from the atmosphere.

Agricultural methane emissions are very large and include both diffuse and discrete sources. They are dominated (67%) by gas produced during fermentation in animal guts and expelled in breath, burps, and farts, or via manure under wet conditions. Methane emissions in breath, burps, and farts might be reduced by up to 80% by mixing algal additives into cattle feed (e.g., Futurefeed, 2019). Regarding manure, cattle pats don't produce much methane under dry conditions, but large manure tanks in industrial farming are major sources. Thus, dealing with manure and emissions from manure tanks can make an important difference. Given that cattle feedlots and barns can be super-emitters, even proportionally small reductions in their emissions can have large impacts on the overall methane budget.

Methane release from wetland rice paddies accounts for about 20% of agricultural methane emissions and might be reduced in theory; for example, through careful management of water levels. But this approach is fraught with problems because it requires advanced skill levels and governance and may lead to reduction of crop yields.

Human-caused biomass burning comprises the remaining 13% of agricultural methane emissions. Methane emissions from biomass burning are important on a global scale, and most relate to the burning of tropical savanna and crop waste, but also of forests. These fires produce methane, carbon monoxide, carbon dioxide, and particulate air pollution. They also remove essential soil nutrients and impoverish soil organic carbon, which is carried away from the region in smoke. While there are ways of limiting these ravages of biomass burning to some extent, there will always be substantial losses. It seems advisable to greatly reduce the practice of biomass burning in agriculture, which lets everything blow away into the atmosphere; a potential alternative is to create biochar and use that for soil re-enrichment.

Biochar consists of compounds that are highly resistant to breakdown and can therefore survive for thousands of years in soils, acting as a stable means of carbon removal and storage, which removes $CO_2$ from the climate system. It can help increase the fertility of acidic soils, increase agricultural productivity, and provide protection against certain crop diseases. Moreover, biochar helps to improve soil structure, properties, and health; to absorb and stabilize pollutants and contaminants; to reduce nitrate leaching into water courses; and—importantly for the topic discussed here—to suppress soil emissions of both methane and nitrous oxide (Shackley et al., 2012).

The promise of biochar addition is well illustrated by observations in the Amazon rainforest. Healthy tropical forest soils contain approximately the same amount of carbon as the lush vegetation above it. Modern changes of Amazonian forest into agricultural land have reduced soil organic carbon contents by 20–50%. However, as much as 5000 to 2500 years ago, the ancient peoples of the Amazon already knew how to transform such impoverished soils, using charcoal. This created dark soils that show clear signs of human interaction, and which contain high amounts of phosphorus and calcium. These soils are known as Amazonian Dark Earth or Terra Preta do Indio, and their productivity remains higher than that of surrounding soils. For photos contrasting them with the normal so-called oxisols of the region, see Glaser et al. (2001). Note that discussion continues as to whether this higher productivity is really caused by the biochar itself or by potential retention of nutrients derived from other plant and animal residues or ash that were mixed into it. Regardless, Terra Preta beautifully illustrates the long timescales over which biochar addition can drive a removal of $CO_2$ from the climate system into soil. For that reason, we will encounter biochar again in Chapter 4.

Finally, dealing with methane from diffuse sources, no matter their origin, is a complicated matter. The gas would need to be extracted from the open atmosphere and then processed. A recent concept suggests capturing methane with chemical catalysts in chambers through which air is pumped by arrays of electric fans powered by renewable energy. One possible way to cost-efficiently deal with the methane would be, though it sounds paradoxical, to convert it into $CO_2$ (Canadell and Jackson, 2019). In terms of climate impact, this makes sense because, molecule for molecule, methane is a much more potent greenhouse gas than carbon dioxide. Because methane exists in much lower concentrations in the atmosphere than $CO_2$, conversion of the entire accumulated amount of human-caused methane in today's atmosphere would add 8.2 billion tons of $CO_2$, which is equivalent to just over 2.5 months of current global anthropogenic $CO_2$ emissions. Yet, because methane is a much more potent greenhouse gas than carbon dioxide, this conversion would still reduce the global warming impact by 15%. And as we saw before, further reduction of the global warming impact could be achieved if the methane, or $CO_2$ after conversion, were to be captured and stored away from the climate system.

* * *

The agricultural use of artificial fertilizer causes release of $CO_2$ during the use of energy to produce the fertilizers and of $N_2O$ when the fertilizers react in the environment. In specific terms, up to 2% of the total global energy consumption is used, causing $CO_2$ emissions, to transform nitrogen gas into reactive nitrogen by means of the so-called Haber Bosch process, which fixes nitrogen with hydrogen to produce ammonia—a critical part in the manufacture of artificial fertilizers. The Haber-Bosch process delivers some 100 million tons of reactive nitrogen per year. Annually, another 45 million tons of nitrogen are fixed naturally, especially in vegetation and to a lesser extent in association with the rearing of animals.

Of the total 145 million tons, only about 25 million tons are consumed in the form of plant and animal protein—about 120 million tons is lost. Though most of this lost nitrogen enters the atmosphere as nitrogen gas, a considerable portion does so as $N_2O$. In fact, agriculture causes about 80% of the total global $N_2O$ human-caused emissions, along with about 40% of the total global human-caused methane emissions. Some 50–70% of the annual 120 million tons of nitrogen loss occurs in the landscape. The next greatest loss happens during the generation of meat protein from plant protein, when major nitrogen release takes place in the form of manure, only half of which is reused as fertilizer. In sum, the entire process of using artificial fertilizer from the Haber Bosch process to generate plant and animal proteins is rather inefficient and wasteful and leads to large $CO_2$, methane, and $N_2O$ emissions.

Because of the major inefficiency of creating animal protein from plant protein—with strong $CO_2$, methane, and $N_2O$ releases along the entire agricultural process chain involved—high levels of meat and dairy consumption are especially substantial contributors to global climate change. This is true in particular for the intensive factory-farm method of meat raising. Meat raising via conventional grazing also has major greenhouse-gas impacts, through land degradation, which also is an important driver of regional desertification. The time has come for development of sound alternatives. There are three main, partial solutions to this problem.

First, a recent special report of the IPCC drew attention to the exceptionally high meat and dairy consumption in western society (IPCC, 2019). It is important to emphasize that these findings do not mean that everyone should stop eating meat—in some places people simply have no other choice. Rather, the report calls for cutting back to reasonable quantities, especially in western society, and states that more people could be fed using less land if individuals cut their meat consumption down to reasonable quantities. This would also be beneficial to public health. Note also that chicken and pork are considerably less carbon-intensive than beef. General food wastage is a closely related issue. The report finds that 8–10% of total global greenhouse gas emissions is associated with food loss and waste. Reduction of food loss and waste, therefore, has to be an important strategy for emissions reduction.

Second, the microbial protein method for producing slow-release organic fertilizers we encountered in section 3.2 can help not only with utilization and thus emissions reduction of industrial $CO_2$, but also with reduction of agricultural $N_2O$ emissions, by reducing the need for artificial fertilizers. Another way of reducing agricultural emissions of both $N_2O$ and methane focuses on the spreading and burial of biochar, as was discussed previously with respect to methane emissions reduction through landfill soil covers.

Third, the greenhouse gas emissions associated with factory-farm and conventional grazing methods of meat raising may be reduced through implementation of regenerative grazing practices. Regenerative grazing allows the production of meat while regenerating the landscape. This greatly reduces the need for artificial fertilizers and reduces greenhouse gas emissions.

* * *

Aside from $N_2O$, other nitrogen oxides also remain a focus for emissions control. These concerns are dominated by $NO_x$ emissions from combustion engines, both in stationary installations and in vehicles. The limitation of $NO_x$ emissions is equally important for engines using fossil fuels and those using biofuels. Society is beginning to feel real consequences from the targets that governments have set for $NO_x$ emissions. For example, to enforce compliance with such targets in The Netherlands, a May 2019 court ruling gave rise to a series of measures, including drastic day-time reduction of speed-limits on all motorways.

The two dominant processes of $NO_x$ formation lead to so-called thermal $NO_x$ and fuel-bound $NO_x$. Thermal $NO_x$ forms at high flame temperatures; its formation increases exponentially with combustion temperature and following the square root of the oxygen quantity in the combustion zone. Fuel-bound $NO_x$ formation depends on the original fuel nitrogen content. In consequence, $NO_x$ formation is most effectively reduced by reducing flame temperature and excess oxygen availability, and by utilization of low-nitrogen fuels. Exhaust gas recirculation into the combustion chamber results in $NO_x$ emissions reduction by reducing oxygen concentrations in the combustion chamber. Injection of water, or anti-detonant, into the combustion chamber with the fuel mixture reduces $NO_x$ emissions by lowering combustion temperatures. Further technologies for reducing $NO_x$ emissions are known as flameless oxidation and staged combustion. Finally, any $NO_x$ that still forms may be (partially) removed using processes such as selective catalytic reduction and selective non-catalytic reduction, which can achieve $NO_x$ removal efficiencies of about 80% and 60%, respectively.

Reducing emissions of the synthetic greenhouse gases (for example, CFCs, HFCs, fluorinated greenhouse gases) depends primarily on legislation to inhibit their production and application. For example, the Montreal Protocol of 1987 banned CFC gases because of the damage they cause to the stratospheric ozone layer, and CFC concentrations have stabilized since then. Meanwhile, however, HFC emissions more than tripled since 1990 in the European Union for two main reasons: first, HFCs replaced ozone-depleting hydrochlorofluorocarbons (HCFCs) in refrigeration and air conditioning; second, the refrigeration and air conditioning sector itself expanded strongly. HFC emissions have been reduced since 2015 both across the European Union and within individual member states as a result of policies and measures designed to reduce leakage from refrigeration and air conditioning equipment, encourage the recovery of gases at the end of equipment lifetime, promote transitions to non-HFC refrigerants, ban the use of HFCs in specific applications, cap HFC supplies, and limit the use of refrigerants with high global warming potentials in new vehicles.

Since the 1950s, $SF_6$ gained prominence for insulation and load breaking in transformers and circuit breakers, most notably in high-voltage circuit breakers. Use of $SF_6$ has enabled the redesign of electrical equipment, making it smaller, easier to maintain, and safer for higher-voltage loads. Ironically, wind turbines, a cornerstone of renewable energy generation, also contain $SF_6$ and can leak some of it to the atmosphere. Any escape of $SF_6$ has severe implications for climate change because $SF_6$ has a 100-year global warming potential that is 23,500 times greater than that of $CO_2$, and survives in the atmosphere for more than 3000 years. Given that $SF_6$ has to be kept above a minimum pressure in switchgear to work effectively, any loss because of leakage must be topped up, fueling further potential leakage. Minimizing $SF_6$ emissions, therefore, is all about regulations and innovations to find replacements and control leakages and losses during maintenance and repairs, as well as losses during equipment construction and decommissioning and during transport of the gas.

$NF_3$, with a global warming potential 16,000 times greater than $CO_2$, is used as a chamber cleaning gas in the manufacture of semiconductors, flat-panel displays such as monitors and TVs, and other electronic devices. Electronic gases such as $NF_3$ are needed in thin-film deposition technologies that are used to make semiconductor and photovoltaic cells. And—here comes the irony—some of the greatest demand for the super-greenhouse gas $NF_3$ now comes from the thin-film solar PV industry, which develops solar panels for the energy transition that aims to limit $CO_2$ emissions reduction. While it is important to undertake detailed, full-lifecycle assessments to determine the net greenhouse warming implications, it clearly makes abundant sense to minimize $NF_3$ losses to the atmosphere, as is true for all fluorinated gases with very high greenhouse warming potentials. Yet, the atmospheric concentrations of $NF_3$ have been increasing very rapidly, and it has an atmospheric lifetime of the order of 500 to 600 years. This $NF_3$ increase therefore represents a substantial and undesirable global warming impact. But at the same time, we need to consider that this impact replaced an even larger impact that would have been suffered if $NF_3$ had not been used to replace other fluorinated gases with even higher global warming potentials or atmospheric lifetimes, such as $SF_6$ or $C_2F_6$. Manufacturers are now increasingly seeking to reduce the use and loss of both $NF_3$ and other fluorinated greenhouse gases in their production processes. Approaches include efficiency increases; switching to alternative gases with lower greenhouse gas potentials, for example from $SF_6$ to $NF_3$ and eventually perhaps to $F_2$; and gas capture and reuse or destruction. Still, the intensive electronics industry in China, Korea, Japan, and Taiwan continues to present major point sources of fluorinated greenhouse gases, including the longest-lived one of all, $CF_4$, with an atmospheric lifetime of 50,000 years.

Measurements for another super-greenhouse gas, $SF_5CF_3$ with a global warming potential 17,500 times greater than $CO_2$, conclusively indicate that atmospheric concentrations grew rapidly from the 1950s and the rate of rise then sharply declined after the late 1990s until it became essentially zero from 2003. The gas was formed during (you best sit down for this) "electro-chemical fluorination for the production of perfluoro-octanyl-sulphonate and other fluoro-surfactants, which were used in the manufacture of foams and stain-resistant coatings" (Sturges et al., 2012). This process was phased out around the year 2000. Although its rate of rise became zero, $SF_5CF_3$ has an atmospheric lifetime of about 900 years, so that atmospheric concentrations have not declined since its emissions were halted.

## 3.4. Emissions Avoidance

In the literature, the term emissions avoidance is loosely used for a wide range of hypothetical scenarios, many of which refer to the potential of adding a smaller, but still positive, amount of greenhouse gas to the atmosphere in replacement for a current process that emits more. The matter is often confused even further by lumping together the physical removal and assumed avoidance of greenhouse gases. Here, I use emissions avoidance for practices that follow the philosophy that no measure of emission control can ever be as good as the measure that prevents or avoids the emission to begin with. As such, emissions avoidance is intimately related to energy efficiency, energy awareness, and active management of society's $CO_2$ and other greenhouse gas footprints. Indeed, emissions avoidance goes beyond just energy and $CO_2$; it applies to all processes that cause large-scale emissions of all greenhouse gases. Emissions avoidance is a field of great innovative activity, and exciting new approaches are emerging every day.

Emissions avoidance needs separate attention because it is a critical player in curtailing climate change, along with emissions reduction and negative emissions (Figure 2.5). The rationale for emissions avoidance as an indispensable part of our strategy in dealing with climate change is straightforward. The Bloomberg New Energy Finance (2019) report presents future projections based on current economic and development trends. It considers that global power demand will grow by 62% between the end of 2018 and 2050 and foresees that, by 2050, the energy mix will comprise 31% fossil fuels and 62% renewables, with nuclear covering the remainder. Fossil fuel in 2050, therefore, is projected at $1.62 \times 0.31 = 0.77$ times the fossil fuel use of 2018, or 77%. So, even though renewables would have gained rapidly in importance within the total energy mix, from about 10% in 2018 to 62% in 2050, there would only be a 23% reduction in fossil fuel use and therefore emissions from 2018 to 2050. This is because the rapid increase in renewables is almost offset by a rapid increase in total energy demand.

The Bloomberg report projects that emissions will be kept constant until 2026 and then drop by about 2% per year. Alarmingly, starting with current emissions of 10 GtC per year, such an outlook translates into emissions of an additional 280 GtC or so by 2050. To keep within the 1.5°C warming of the Paris Agreement, we only have a window of about 200 GtC emissions starting from 2017 values, or about 400 GtC to keep within 2°C warming (Goodwin et al., 2018). The Bloomberg projections would see the emission limit for the 1.5°C warming target being broken as early as 2039 and that for the 2°C target by about 2075.

These simple calculations illustrate the absolute urgency of curbing the growth in total energy demand (that is, to practice large-scale emissions avoidance) to optimize the impacts of the energy transition to renewables. Incidentally, the calculations also illustrate in the most straightforward way that the emissions limits for the Paris Agreement's warming targets are very likely to be exceeded well before 2100, which means that society should in addition urgently prepare for large-scale greenhouse gas removal from the climate system (see Chapter 4).

One of the main routes for new emissions avoidance is through installa-tion of equipment in new power plants and factories to capture $CO_2$ emis-sions before they enter the atmosphere and then storing the captured $CO_2$ in long-term storage facilities (Figure 3.10). This is called Carbon Capture and Storage (CCS), or carbon sequestration. This approach crosses over into the emissions reduction territory as well, because existing power plants and factories can be retrofitted with CCS options. If executed perfectly, which is not likely, CCS could in theory make a facility carbon neutral, which means that it has zero emissions. In practice, there are always some losses, leakages, and emissions made during processing. But facilities can be built using CCS to a point where they have very low emissions relative to what they would have without the CCS option. A remaining key question is whether building carbon-capture-enabled fossil fuel power generation still makes sense, eco-nomically, given that building new renewable energy facilities has become cheaper than running existing coal facilities (Mahajan, 2018; Gillingham, 2019; see also section 2.5).

There are four main geological storage types used in CCS (e.g., Zhang and Song, 2014). First is structural or hydrodynamic trapping: supercritical $CO_2$ injected in a reservoir is then trapped under an impermeable layer of so-called cap rock such as a clay or salt formation. Second is so-called residual trapping, which uses porous reservoir rocks as a sort of rigid sponge: super-critical $CO_2$ injected into the reservoir displaces fluid, and as the $CO_2$ moves on the fluid moves back and traps $CO_2$ in the pores. Third is solubility trap-ping, which relies on the fact that $CO_2$ dissolves into the salty reservoir fluids, which slightly increases the density of the reservoir fluid, causing the

CARBON SEQUESTRATION OPTIONS

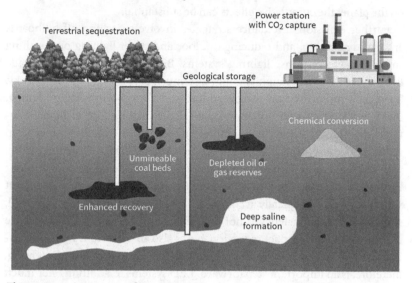

**Figure 3.10** $CO_2$ capture from power plants or industry and long-term storage. Also shown is terrestrial sequestration, but this is not relevant yet to the argument made here; it is addressed in Chapter 4.

*Source*: Following Appel (2019). https://towardsdatascience.com/tackling-climate-change-with-machine-learning-78d1e185b3ec

$CO_2$-enriched fluid to sink to the bottom, thus trapping the $CO_2$. Finally, there is mineral trapping, where $CO_2$ is dissolved into reservoir fluids and then over time reacts with the rock to form carbonate minerals inside the reservoir. CCS will again be discussed in Chapter 4.

A wide range of emissions avoidance measures will be familiar from everyday life. Common examples include energy efficiency measures such as flying less and avoiding unnecessary driving by walking, cycling, using public transport, or teleconferencing; lowering heating settings and avoiding excessive cooling with air conditioning; switching off equipment that is not in use, rather than keeping it on standby; conserving water; recycling products that have resource- and energy-hungry manufacturing pathways, such as paper, plastics, metal, glass, and batteries; repairing and updating equipment rather than discarding it; and eating more fresh, seasonal, and regional vegetables that require less transport. All of this might appear to yield only small benefits when viewed per person, but with more than 7 billion people on the planet the cumulative effects can be substantial.

Further emissions avoidance strategies involve switching off billboards and facade lighting, and reducing outdoor and street lighting or installing smart demand-response lighting systems. Billboard and especially facade lighting result in major ambient light loss, much of it directly up into the sky. Together with losses from street lighting and always-on building lighting, this is not only wasteful of energy but also a major cause of light pollution, which disturbs bird, bat, and insect behavior, as well as human sleep rhythms. Because street lighting may represent an important safety measure, with some studies reporting notable reductions in crime rates under outdoor lighting, plain "switching it off" does not seem to be the best solution. But still, street lighting typically accounts for 25–50% of a municipal energy bill, so that improved efficiency is not only interesting for emissions avoidance and environmental impact reduction, but is also an important value proposition. Transitioning from traditional lighting to highly efficient LED lighting, therefore, is an important step forward. LED systems in addition offer much longer lifespans, which reduces both maintenance and waste.

There is increasing interest in the construction sector in developing highly efficient, energy-neutral buildings, both in terms of the construction process itself and in terms of building operation, which includes heating, cooling, water management, and waste recycling. We are all familiar to some extent with the energy reductions that can be achieved by properly insulating buildings against excessive heat loss or warming and with the increasing trend of incorporating technological innovations such as smart glass and smart thermostats. In addition, an entire discipline, known as bioclimatic design, focuses on how to optimize building styles for their specific environments. The bioclimatic design concept emerged around the 1950s and has gained interest especially since the 1990s. Initially, its main aim was to create comfortable buildings through an understanding of the microclimate they are placed in and then implementing suitable natural ventilation and lighting, and passive heating and cooling. It has since evolved to include also landscape preservation and integration of buildings within the landscape (for example, earth sheltering), water management, waste recycling, the use of sustainable building materials and methods, and optimized inclusion of adaptive smart technology and materials. Today, ambitions even reach beyond energy-neutrality toward energy-positive developments, which means that buildings produce more energy than they use. An example is Snøhetta's initiative for the energy positive Svart hotel within the Arctic Circle in Norway.

Broader measures of emissions avoidance include nature conservation and restoration. Examples are minimizing deforestation and forest degradation (for a dramatic quantification, see Maxwell et al., 2019) and replanting deforested regions; management of savanna burning; preservation and restoration of peatlands; prevention of desertification; and avoiding the destruction of productive coastal environments and re-establishing coastal systems such as mangroves, mudflats, and seagrass ecosystems. Other approaches focus on controlling leakage of greenhouse gases into the atmosphere, through capture and utilization of gas (methane) emissions from landfills and wastewater processing; and methane flare control and enhanced leak-proofing of petroleum-industry infrastructure.

Next, we get to a category of emissions avoidance measures that has wider significance, and which was highlighted in a recent IPCC special report on climate change, desertification, land degradation, sustainable land management, food security, and greenhouse gas fluxes in terrestrial ecosystems (IPCC, 2019). It concerns limiting food wastage, which today amounts to about one third of all food produced. Limiting food wastage would not only avoid unnecessary emissions but could also go a long way to reducing world hunger, inequality, and conflict. So, limiting food wastage even makes abundant sense on that basis alone. The same IPCC report also suggests that major emissions reductions could be achieved, relative to the present state of affairs, through a change in humanity's diet with specific focus on reducing excessive meat consumption in favor of a more balanced diet, which would benefit population health as well (see section 3.3). Moreover, as future food production needs to rise to feed the rapidly growing world population, such a change in diet represents a major emissions avoidance measure.

And finally, microbes again. Microbes can help not only in processes for emissions reduction, but also in processes for emissions avoidance. Implementation of more sustainable and efficient pathways of nitrogen conversion into edible protein using microbes can avoid a large proportion of the $CO_2$ emissions associated with the Haber Bosch process and a major portion of the large $N_2O$ emissions associated with the annual 120 million tons of nitrogen loss in the agricultural process chain. This will reduce the requirement for new fertilizer production, which avoids $CO_2$ emissions, and more nitrogen will be efficiently recycled, which avoids $N_2O$ emissions. One approach, for example, achieves these goals by upgrading waste treatment plants for the use of microbes, which combine waste-derived nitrogen with waste-derived methane or $CO_2$ to generate microbial protein, at almost 100% efficiency. The microbial protein can be used in animal or human food directly, or for other purposes. Through reduction of the need for animal protein generation from plant protein, this approach also has potential for reducing agriculture's methane emissions.

Independent from the microbe-based approaches, other solutions for improving nitrogen usage and thus avoiding or reducing $N_2O$ emissions consider improvements in the efficiencies of fertilizer application and animal rearing. There is profound interest, for example, in so-called enhanced-efficiency fertilizers, which combine fertilizer with breakdown inhibitors. These can increase nitrogen uptake by plants and so reduce nutrient loss through leaching, runoff, and gas emissions. Other research focuses on regulating the timing and amount of irrigation, given that $N_2O$ is often produced in greater quantities when soils are wetter. In addition, the timing of fertilizer application matters. Fertilizer applied within weeks after planting rather than during or before planting increases the actual use of nitrogen in the plants, instead of being lost to groundwater or the atmosphere. Precision fertilization close to the crop root zone also helps to reduce nitrogen loss.

With respect to $NO_x$ and the synthetic greenhouse gases, removing the causes of emissions is key to emissions reduction, while not expanding the processes that cause the emissions is key to emissions avoidance. For $NO_x$, emissions avoidance thus means not continuing practice as usual and instead ensuring that the number of planned new installations is reduced. Also, any new installations that will proceed should not be conventional types, but state-of-the-art types that are optimized for low $NO_x$ emissions by, for example, low-nitrogen fuel use, exhaust gas recirculation, anti-detonant injection, flameless oxidation, staged combustion, and selective catalytic or non-catalytic reduction. For the synthetic greenhouse gases, emissions avoidance also means not expanding the applications that use them, or at least using them in new installations that are optimized for destroying any escaping gases, and investing substantially in development and implementation of less harmful alternative methods. Retrofitting some existing installations with means to destroy escaping gases may also be feasible, though this would fit better under the term emissions reduction.

But, against all of this, in terms of improving energy efficiency for emissions avoidance, stand two gargantuan challenges. First, the world population is growing rapidly, and the average level of technological development per capita is increasing too. As a result, there is a rapid increase in the demand for energy: more and more is needed. This challenge can only be met without massive emissions growth by helping developing nations skip as much as possible the fossil fuel addiction that accompanied developments in western and eastern societies. Direct implementation of widespread renewable energy is essential to prevent population growth and technological development driving the demand for fossil fuel energy beyond control. Strong international support is needed for developing societies to implement this model of skipping of the "fossil fuel energy generation" in favor of direct adoption of a renewable energy. Such a move would have additional benefits for public health in the developing regions, by avoiding pollution from cheap, out-of-date, fossil fuel technologies.

The other major challenge in improving global energy efficiency is of a more insidious nature. This concerns the cryptocurrency hype. When bitcoin started in January 2009, it did not demand much computing power, and thus energy, but with its exponential expansion—along with other cryptocurrencies—the demanded computing power started to pose enormous energy demands. By April 2017, bitcoin consumed an annual equivalent of 11–11.5 TWh of energy. And then its energy demand exploded beyond belief. An historic maximum annual equivalent of 75 TWh was reached in September–November 2018. Thereafter, the bitcoin value temporarily crashed. The bitcoin energy consumption tracker at Digiconomist indicates that consumption stood at an annual equivalent of 66.7 TWh in June 2019, which still equates to the domestic energy consumption of the Czech Republic, a country of 10.6 million people. Since then, consumption has been increasing rapidly again.

There are (contested) claims that bitcoin is running on 75% renewable energy, but this is somewhat beside the point. Bitcoin, and other cryptocurrencies, represent an enormous additional growth in global energy demand, and this expanded energy use could and should have been avoided. Because of the astronomical growth in energy demand by cryptocurrencies, the renewable energy consumed is not used to offset fossil fuel energy. Plainly speaking, massive energy-use expansion that is partially or even totally covered by renewables does *nothing* to reduce $CO_2$ emissions. In addition, there are increasing concerns about the electronic waste created by the cryptocurrency hype. The cryptocurrency mining devices commonly serve only less than two years and cannot be repurposed for other uses. This represents another great pressure on resources, and another major source of emissions, during manufacturing and decommissioning.

## 3.5.  Downsides and Challenges

Theoretically, it might seem easy to compare different energy sources and to then reach a verdict about which one is cleanest, or cheapest, or both. Reality, however, is never easy. All large-scale industrial processes have downsides and challenges that need to be overcome before or during their implementation; fossil fuels have them, and so do renewables. We need to be aware of these downsides and challenges, and account for them, to ensure that we don't simply swap one evil for another. We especially need to know early on whether, and how, any new downsides may be minimized by a sound framework of awareness, planning, monitoring, and control that is implemented right from the get-go. This will help with introducing new systems that have the least detrimental impact.

But making impact assessments and comparisons between industrial processes is not trivial—all sorts of biases can be introduced. For example, the world is full of fossil-fuel infrastructure and support already. It is therefore easy to end up unfairly comparing the detrimental impacts of infrastructure construction requirements for renewables with a perceived absence of such requirements for fossil fuels. The existence of fossil-fuel exploration and exploitation structures, pipelines, fleets of tankers, an abundance of large-scale refineries and storage facilities, and a mature distribution network should be fully accounted for alongside any new fossil-fuel infrastructure when making comparisons with the infrastructure needs of renewable energy sources. Why? Because indiscriminately comparing the need for new hydrogen refilling stations with the abundance of existing petrol refilling stations would lead to a false case in favor of petrol rather than hydrogen. The one-off investment for installing new hydrogen refilling stations should instead only be compared with a similar one-off investment for installing petrol refilling stations. Only once this is accounted for can the operational costs of the two fuels be compared on a fair, equivalent basis.

A similar argument applies to the costs of pollution in terms of climate change and public health impacts. These costs need to be equally and fairly accounted for with respect to any energy source considered—either per kilowatt of energy produced or, in the case of transport, per person (or freight tonnage) over a standard unit of travel distance. Because the fossil-fuel sector has so far largely escaped dealing with these issues, the May 2019 working paper of the IMF considered that it has been heavily subsidized over the years (section 2.5). To fairly operate on a level playing field, renewables need to be viewed in exactly the same way as fossil fuels; both either fully unsubsidized or subsidized to the same level. These sorts of issues call for very complex evaluations on a case-by-case basis. Because of these complexities, we will not compare different systems with each other in this section, but merely highlight some key drawbacks and challenges reported for each system by itself.

Solar, wind, and hydropower utilities have expanded massively and rapidly. This expansion has revealed several hurdles and limitations that need to be overcome. One issue concerns space. Large, utility-scale solar energy systems require 1–4 hectares per Megawatt. This creates potential competition with land surface needs for habitation, agriculture and food production, or biodiversity and conservation. Removal of trees and bushes, to reduce shading, also impacts negatively on biodiversity and conservation. To resolve such conflicts (on land at least), many nations are investigating floating solar power plants. This also avoids social costs, including land requisition problems. Other solutions may come from innovative approaches, such as solar-wind energy towers, which require less than a tenth of the surface area for similar output capacity.

Wind energy facilities have a typical space requirement of 0.5–1 hectare per Megawatt. Many nations have already implemented extensive offshore wind farms to reduce conflict with other land uses and to alleviate social resistance to extensive new onshore wind projects. In addition, there is a major trend of installing facilities that comprise individual wind turbines of very high capacity, to optimize the amount of energy gained per unit area. The largest turbine so far is the GE Haliade-X, which is being constructed outside Rotterdam, Netherlands. At 260 meters, it is almost three times as high as the Statue of Liberty. It has a capacity of 12 Megawatts, and a sensational capacity factor of 63%. For scale, powering a typical household for a day with a Haliade-X would require less than one revolution of its gigantic fan. One

Haliade-X can power 16,000 households based on typical wind conditions for an average North Sea site.

A second issue with increasing reliance on wind and solar power arises from intermittency in the power delivery, which we encountered before when discussing the capacity factors of such systems. The intermittency can be addressed by improved energy storage, which we also saw before as key to maximizing the capacity factor of solar CSP systems. On solar PV systems, storage could come in the form of large-scale battery facilities, such as that installed by Tesla in South Australia. Alternatively, it can be achieved through a pumped-hydro approach, in which energy is used to pump water to higher reservoirs when available, allowing it to be harvested in the form of hydropower when needed. In wind energy systems, technological improvements are widening the window of operational conditions, both by transitioning to giant wind turbines with larger fan blades and by design improvements. Yet another method relies on multitudes of household and/or car batteries linked into a distributed grid system. On a higher level, it is becoming evident that interconnected so-called super grids will have to be created over wide geographic areas to ensure that lack of renewable energy production in some sectors of the grid can be offset by ongoing production in other sectors. A much-discussed example is the Europe-wide super grid, through which Europe may achieve a 90% renewable energy level by about 2040.

Hydropower is a major success story in terms of energy generation. But, like all industrial infrastructure, it does not come without sacrifices. First, stored and pumped hydropower rely on reservoirs, which are often constructed by flooding valleys behind major dams. This has major impacts on river-flow characteristics both upstream and downstream of the dam and reservoir, and thus affects sediment transport, fish migration, riverine ecosystems and biodiversity, and so on. The flooding of valleys in addition drowns land-based ecosystems and biodiversity, often also including human habitation, agricultural lands, and cultural heritage. Moreover, creation of new reservoirs is a major geotechnical effort, and hazards need to be controlled. For instance, large-scale flooding of previously dry valleys alters hydrostatic pressure gradients in sediment layers, which can cause slope instability and collapses into the lake, with potentially catastrophic consequences such as the 1963 Vajont Dam disaster in Italy.

Finally, questions remain about the actual emissions reduction that can be achieved by switching from fossil fuels to hydropower involving dams and reservoirs. This is especially important in warm regions. There, major methane production occurs both from decomposition of the vegetation drowned upon creation of the reservoir and from vegetation and algae that grow, die, and decay during lake-level fluctuations. The methane is formed in particular on the lake floor and within lake-floor sediments and is released when the overhead pressure drops when the lake level is lowered. It is a major cause for concern because methane has a much higher global warming potential than $CO_2$. A study of hundreds of dams in the Mekong River basin has found that the $CO_2$-equivalent greenhouse gas emissions of one in five of the dammed reservoirs are equal to, or higher than, those from theoretical fossil-fuel burning to achieve the same energy production (Räsänen et al., 2018). Similar concerns apply to reservoirs worldwide (Deemer et al., 2016), and this doesn't even include all of the pathways that contribute to net emissions counting, such as concrete production and transport, or the termination of $CO_2$ absorption by the—now flooded and dead—original vegetation. From a climate perspective, therefore, hydropower should be planned carefully to maximize energy production and minimize reservoir surface area. Evidently, hydropower developments cannot be simply assumed to be a zero-emissions process. Optimization of results depends on full-cycle greenhouse gas analysis, region/setting-specific planning, installation (by preference after vegetation removal), monitoring, and regulation for net greenhouse-gas emissions.

Tidal barrage developments come with their own broad suite of environmental issues. Any barrage (dam) affects water flow into and out of the estuary and thus the highly productive estuarine habitats both in the water—which fluctuates from marine, through brackish, to fresh—and in the rich mudflats and salt marshes. These habitats are extremely important feeding grounds for fish, marine mammals such as seals and porpoises, and birds. They also have great value to humans because estuaries are prime areas for fishing and shellfish industries. In addition, healthy estuarine environments and other shallow-water environments are important places for organic carbon deposition into sediments and thus net carbon drawdown from the climate system. In consequence, plans for barrages, such as that discussed for the Severn estuary, always meet with important (and justified) environmental concerns, and the emissions reduction achieved from the actual tidal power generation would need to be contrasted with the reduction of net carbon drawdown into the sediments to get a view of the full life-cycle carbon impact. And, again, each assessment will need to be region and climate-zone specific.

With respect to OTEC applications, most concerns are focused on the very large volumes of deep water that need to be pumped toward the surface in open OTEC systems. The concerns arise because higher-pressure and colder deep waters contain elevated dissolved $CO_2$ concentrations, relative to lower-pressure and warmer surface waters. This leads to $CO_2$ outgassing when the deeper waters are depressurized and warmed up when brought up toward the surface in the working-fluid-cooling or vapor-cooling stage. Moreover, deeper waters contain elevated levels of nutrients that, when released near the sun-lit surface, will drive increased productivity. This productivity enhancement might be positively utilized in—for example—aquaculture, but it also has a significant potential to upset regional ecosystems. While initial assessments suggest that the ecosystem impacts might be limited by discharging waters at depths of around 100 meters, rather than at shallower levels, there is some additional energy cost involved in pumping the waters back down to those depths. Hence, it is critical that sound, comprehensive life-cycle studies are performed. These will especially need to compare the emissions reduction achieved from the actual net OTEC power generation with any $CO_2$ outgassing from deeper waters pumped up toward the surface.

Geothermal plants require careful assessments as well. And these assessments again need to be performed on a per-case basis because geothermal plants are not all the same. Geothermal plants use fluids that are pumped through deep underground layers, where they are heated, and then brought back up to the surface. When pumped back up, these fluids contain variable quantities of $CO_2$ and (much less) hydrogen sulfide. The gas quantities reflect the geological conditions of the reservoir that the fluids have passed through to get heated. In current installations, some of the gases may be captured and re-injected back into the reservoir, but most are released and vented through a cooling tower, and some 90% of these gases is $CO_2$. Thus, geothermal power plants still emit some $CO_2$, albeit much less than oil, gas, or coal-fired plants (Table 3.4). However, some geothermal regions can emit much greater quantities of $CO_2$ because of their specific geological characteristics. For example, Turkish geothermal fields reach values of 450–1450 g $CO_2$/kWh (Aksoy et al., 2015). Geothermal plants in these fields, therefore, risk emitting just as much $CO_2$ as fossil-fuel power plants would for the same energy generation.

Near-zero $CO_2$ emissions can be approached in geothermal power generation using so-called closed-loop binary-cycle power plants, which pass the extracted geothermal fluid through a heat exchanger and then completely inject it back underground again. This type of plant is seen as key to future expansion of geothermal power. Low-temperature direct geothermal heating also causes negligible $CO_2$ emissions. Overall, it appears that an optimized low-emissions approach using geothermal energy requires careful geological assessment and greenhouse gas emission monitoring, as well as further improvement and implementation of closed systems. As we saw with hydropower in warm regions, it is a mistake to make a priori assumptions that such renewable energy sources are zero-emissions sources by definition. Sound investigations are needed to reveal the basic operational parameters on a case-by-case basis.

In transport, similar to the well-to-propeller case that we discussed for shipping, complete well-to-combustion and well-to-drive life-cycle assessments are needed for all other fossil and renewable fuel applications, including aviation. For biofuels, such comprehensive analyses need to account in full for the impacts of feed-stock plantations, harvesting, and transport. Doing so reveals issues of scale regarding the potential use of modern biofuels to replace fossil fuels. To support such a transition, vast biofuel feedstock crops are needed, but switching to biofuel crop development at these scales is highly detrimental to the environment and to food production. Notably, land, water, and nutrient requirements for fuel crops compete with similar requirements for both food-producing crops and pristine or reestablished wilderness. Economic pressures tend to favor rapid replacement

**Table 3.4** $CO_2$ emissions per kWh for different methods of power generation.

| Method | $CO_2$ emissions (g $CO_2$/kWh) |
|---|---|
| Geothermal power | 0–400 |
| Oil-fired power | 700–900 |
| Gas-fired power | 450–1250 |
| Coal-fired power | 850–1300 |

*Source*: Kristmannsdóttir and Ármannsson (2003).

of less lucrative food crops and wilderness/biodiversity by lucrative fuel-crop monocultures, leading to food shortages and widespread deforestation. Moreover, the intensive agriculture needed for growing vast biofuel feedstocks is associated with extensive fertilizer use and methane and $N_2O$ emissions. Full life-cycle assessments—from land clearance through farming and harvesting to transport and processing—therefore need to span the entire economic, environmental, and greenhouse gas spectrum. The listed problems were particularly pronounced with first generation biofuels and came to light rapidly as bioethanol use first became popular.

Second generation biofuels may not compete with the food cycle, but still compete for resources, and also require considerably more extraction, thermal, chemical, biochemical, and purification processing than first generation biofuels. These additional steps require energy and thus have considerable impacts on the full-cycle $CO_2$ budget associated with second generation biofuel generation. Third generation biofuels are less affected by these limitations. In addition, the third generation's algal systems can be implemented on land that is unsuitable for agriculture and can use non-fresh water sources such as remediated wastewater and brackish or salt water. Yet, a recent assessment suggests that algal biofuel generation is neither commercially nor environmentally sustainable, mainly because of biological inefficiency (Kenny and Flynn, 2017; Flynn, 2017). The study implied that providing just 10% of EU transport fuels would require ponds three times the surface area of Belgium and that these systems would require fertilizer equivalent to 50% of the current total annual EU food crop needs. Making progress out of this stalemate may depend on genetic modification of algae to make the process more efficient; this is referred to as the fourth generation approach to biofuels. But that step leads to safety concerns in case newly engineered super-algae escape into the natural environment.

For both transport and baseload power provisions, batteries are rapidly gaining a prominent position. Yet, production of batteries is not without its own environmental and emissions costs. These costs come into play especially because current generations still experience aging, which means that only a finite number of charge-discharge cycles may be expected over the battery's lifespan. This problem will diminish as batteries are improved to live longer—more than 5000 cycles are achieved already in modern automotive batteries. Another issue is that current battery generations still have considerable mass, size, and price relative to the amount of charge held, which imposes range and installation limitations. For scale, the energy-by-mass ratio for gasoline is 60 times that of a modern lithium-ion battery, and 34 times that of the typical automotive nickel-manganese-cobalt battery. Countering this, to some extent, lithium-ion power offers some 3 to 4 times the energy efficiency of an internal combustion engine. Batteries furthermore suffer from slow refueling times, given that an 80% recharge typically takes at least 1 hour, and a full recharge takes 1 to 3 hours. Fast-charging technology is developing rapidly, but still remains far from the refueling timescale of the order of minutes offered by more conventional fuels. Finally, batteries require elements such as cobalt, the mining of which poses serious environmental and humanitarian challenges, and requires careful recycling to avoid environmental pollution with their hazardous contents. Intensive ongoing research and development aims to improve the energy-by-mass ratio and reduce the charging limitations and environmental impacts of battery technology. For example, there is a strong push toward development of batteries that use less rare and harmful materials.

Hydrogen-fueled transport offers longer range and faster refueling than electrical transport solutions. But there are significant drawbacks to hydrogen-fueled transport as well. First, hydrogen fuel cells are expensive to produce because they require rare substances such as platinum as a catalyst. Second, greater infrastructure investments are needed to support hydrogen-based transport, relative to electrical transport. Third, there are no geological reservoirs of hydrogen, which means that it must be produced, either as hydrogen gas or as ammonia. Thus, the hydrogen fuel for transport represents an intermediate step in the energy chain, which adds significant fuel costs. In contrast, electrical vehicles use the electricity directly, cutting out the costly intermediate step. Romm (2009) pointed out that a hydrogen fuel-cell vehicle's overall efficiency of renewable electricity use is about 20–25%, while that of a battery-powered electric vehicle is about 80%. This is a major barrier to overcome. One interesting possibility is to concentrate on hydrogen fuel cells more as a replacement for fossil-fuel engines in hybrid electrical vehicles, to provide range extension and back-up fuel—but biofuels are in competition for that role too.

Especially in shipping, hydrogen suffers from the fact that it has a low energy density, so that large storage volumes are needed to achieve the amount of energy needed. This may prevent hydrogen from being used directly in international deep-sea shipping. Its future in shipping may lie more in indirect, synthetic-fuel or ammonia based, roles. Biofuels have specific drawbacks too, especially when not upgraded (hydrogenated). Depending on their source, for example, there may be issues of stability during storage, acidity, lack of water-shedding that may result in biological growth within the fuel tank, formation of waxes, blocking of filters, and increased engine deposits. These issues can stall an engine, which might put a ship at critical risk. To avoid these issues, careful selection of fuels is needed, as well as suitable adaptations to the engine. Contamination of biofuels with water is also a matter of concern because biofuels are susceptible to biofouling. Upgrading of first-generation biofuels in a refinery yields high-quality fuel that avoids the aforementioned issues, but the refining requires energy and thus results in more emissions. Again, full-cycle greenhouse gas assessments are essential.

Use of hydrogen in steel production may negatively affect the processing characteristics and service performance of steel products. Under some conditions, hydrogen can be rejected during solidification of the steel and cause pinhole formation and porosity, which can then cause hairline cracks, embrittlement, blistering, and loss of capacity to resist failure when stretched. Such issues are not limited to steel but similarly affect other metals produced using hydrogen in an effort to cut $CO_2$ emissions. Thus, focused development is needed of new procedures for alloy design that account for detrimental hydrogen influences. In the EW approach to steel making, the key reduction stage can become carbon free if renewable electricity is used, but some $CO_2$ emissions would remain associated with final refining steps to attain the desired carbon content of the final steel product. Overall, analyses across a range of steel-making approaches indicate that only the most ambitious steel-industry development scenarios are conducive to an 80% emission reduction before 2050 (Fischedick et al, 2014).

In the construction and engineering sector, many alternative binders for concrete and alternative building materials are perfectly suitable for replacing traditional cement/concrete in many applications. But major infrastructure projects such as dams, bridges, and major tunnels will for a considerable time continue to rely on traditional concrete for combined reasons of cost, production volume, and material properties and longevity. Through its lifetime, and especially after demolition, conventional concrete (re)absorbs $CO_2$, and some initiatives are evaluating how to quantify this so it can be accounted for in a nation's $CO_2$ budget. This process makes the *net* contribution of the cement/concrete industry to total global emissions smaller than the 8% stated before, which helps in understanding its total impact. Still, let's be very clear that, in reality, such a take on the matter is an accounting fix only, which should never be allowed to suppress the drive toward truly reducing the construction industry's climate change impacts, by transitioning to low-, zero-, or negative-carbon materials.

* * *

Overall, one common message emerges: we cannot let the transition to more sustainable, renewable resources in any major industrial application be guided exclusively by economic arguments. Although it is important to find and implement the more cost-effective solutions, it is equally important that detailed "cradle-to-grave" life-cycle assessments are undertaken, both for emissions and for wider impacts, such as conflicts with other essential services (for example, food production). Such life-cycle assessments are the only means to safeguard against causing irreparable damage while transitioning toward more sustainable, renewable resources. Humanity has unwittingly— and since the 1980s knowingly—created widespread collateral damage during its previous, fossil-fuel-based societal transformation. We should not again blunder mindlessly into a similar situation. Therefore, it is essential that careful, independent research, development, innovation, and continued monitoring and assessment are conducted throughout the upcoming energy transition. Assessments also need to include the impacts of changing climate conditions on the energy system in general and renewable energy potential in particular (e.g., Cronin et al., 2018; Solaun and Cerdá, 2019).

Across all agricultural emissions and including also the use of biofilters in other emission sources, there are concerns that methane emissions reduction may inadvertently increase emissions of other greenhouse gases, notably of $N_2O$. Given that $N_2O$ has a much higher global warming potential than methane, this would entail a seriously detrimental overall impact on climate change. Careful case-by-case assessments are needed, and advanced development is needed of methane treatment methods that do not increase emissions of other greenhouse gases.

We have seen that biomass burning causes substantial emissions of methane, carbon dioxide, carbon monoxide, and particulate air pollution, and that reduction of biomass burning therefore is advisable. Note, however, that this argument strictly applies to agricultural biomass burning. While it is also true in the strictest sense for hazard-reduction burning for fire-season preparations, that practice of biomass burning requires a different balance of assessment with an overwhelming weight on the safety of society, infrastructure, wildlife, and the crews who would be fighting any wildfires.

In addition, efforts in the agricultural sector to reduce emissions or other climate-change impacts, for example because of land-use changes, carry a risk of impacting on both food availability and rural livelihoods. This may happen because of re-allocation of land from food crops to fuel feedstock crops as we discussed under biofuels or because of limitations to land availability for agricultural expansion due to the need to avoid conversion of high carbon landscapes such as forests. The latter is an example of the requirement to practice emissions avoidance wherever we can, as discussed in section 3.4. Other reasons for impacts on food availability are a potential shift to practices that are less greenhouse gas intensive, away from more intensive practices such as dairy and beef production. And finally, reduced-greenhouse gas practices may directly or indirectly affect product prices and the quantity of food production directly through, for example, reduced fertilizer application, or reduced livestock density, and indirectly through, for example, increased production costs (Frank et al., 2017).

## Key Sources and Further Reading

Agence France-Presse. Germany launches world's first hydrogen-powered train. *The Guardian–Environment*, September 18, 2018. https://www.theguardian.com/environment/2018/sep/17/germany-launches-worlds-first-hydrogen-powered-train

Aksoy, N., et al. $CO_2$ emission from geothermal power plants in Turkey. *Proceedings World Geothermal Congress 2015 Melbourne, Australia*. April 19–25, 2015. https://pangea.stanford.edu/ERE/db/WGC/papers/WGC/2015/02065.pdf

Alsharif, M., Kim, J., and Kim, J.H. Opportunities and challenges of solar and wind energy in South Korea: a review. *Sustainability*, *10*, 1822, 2018. https://www.mdpi.com/2071-1050/10/6/1822/pdf

Anderson, S. Regenerative agriculture can make farmers stewards of the land again. *The Conversation*, February 11, 2019. https://theconversation.com/regenerative-agriculture-can-make-farmers-stewards-of-the-land-again-110570

Appel, T. Tackling climate change with machine learning. *Towards Data Science*, October 14, 2019. https://towardsdatascience.com/tackling-climate-change-with-machine-learning-78d1e185b3ec

Architecture 2030. *Insulation. Carbon smart materials palette.* 2018. https://materialspalette.org/insulation/

Argonne National Laboratory. *Life cycle analysis of alternative aviation fuels in GREET. ANL/ESD/12-8*, US Department of Energy, Oak Ridge, Tennessee, June 2012. http://www.energiasustentables.com.ar/bioenergia- en/fundamentos%20adicionales-en.html

Ariizumi, R. Renewable hydrogen: key to a new civilization. *Our World, Science & Technology*, November 1, 2010. https://ourworld.unu.edu/en/renewable-hydrogen-key-to-a-new-civilization

Arnold, T., et al. Nitrogen trifluoride global emissions estimated from updated atmospheric measurements. *Proceedings of the National Academy of Sciences of the U.S.A., 110*, 2029–2034, 2013. https://pdfs.semanticscholar.org/2259/be3e22b2c764f48add754825c54461279f52.pdf

Arnold, T., et al. Inverse modelling of $CF_4$ and $NF_3$ emissions in East Asia. *Atmospheric Chemistry and Physics, 18*, 13305–13320, 2018. https://www.atmos-chem-phys.net/18/13305/2018/acp-18-13305-2018.pdf

Askja Energy. *World's largest electricity producer per capita.* June 4, 2012. https://askjaenergy.com/2012/06/04/worlds-largest-electricity-producer-per-capita/

Ayerm, N.W., and Dias, G. Supplying renewable energy for Canadian cement production: life cycle assessment of bioenergy from forest harvest residues using mobile fast pyrolysis units. *Journal of Cleaner Production, 175*, 237–250, 2018. https://www.sciencedirect.com/science/article/pii/S0959652617329700

Bakhtov, A., et al. Alternative fuels for shipping in the Baltic Sea region. HELCOM-Helsinki Commission, March, 2019. http://www.helcom.fi/Lists/Publications/HELCOM-EnviSUM-Alternative-fuels-for-shipping.pdf

Baker-Munton, C. Dead in the water. *Southeast Asia Globe*, June 5, 2019. https://southeastasiaglobe.com/dead-in-the-water/?fbclid=IwAR3hQ3Tgje7736bagZZWFgUjloyEJBC-qSfeccsgUbz_CXKM1_ZXWxcPCfI&sfns=xmo

Bannon, E. Battery, hydrogen and ammonia-powered ships by far the most efficient way to decarbonise the sector—analysis. *Transport and Environment*, November 15, 2018. https://www.transportenvironment.org/press/battery-hydrogen-and-ammonia-powered-ships-far-most-efficient-way-decarbonise-sector--

Barrera, O., et al. Understanding and mitigating hydrogen embrittlement of steels: a review of experimental, modelling and design progress from atomistic to continuum. *Journal of Material Science, 53*, 6251–6290, 2018. https://link.springer.com/content/pdf/10.1007%2Fs10853-017-1978-5.pdf

Batarags, L. This futuristic hotel is going to be built at the base of a glacier in remote, northern Norway—and it looks like it's straight out of a sci-fi movie. *Business Insider*, October 20, 2018. https://www.businessinsider.com/svart-hotel-norway-glacier-energy-positive-arctic-circle-adventure-tourism-2018-10?r=AU&IR=T

Battery University. *BU-104a: comparing the battery with other power sources*. March 28, 2019. https://batteryuniversity.com/learn/article/comparing_the_battery_with_other_power_sources

Biochar.org. *Slash and char: soil charcoal amendments (agrichar or biochar) maintain soil fertility and establish a carbon sink*, Cuvillier Verlag, Göttingen, 190 pp., 2007. http://www.biochar.org/joomla/index2.php?option=com_content&do_pdf=1&id=24

Bloomberg New Energy Finance. New energy outlook 2019. *Bloomberg Finance LP*, 2019. https://about.bnef.com/new-energy-outlook/

Brown, J. Alternative fuels: the future of zero emissions shipping. *Environmental Defense Fund Europe News*, July 31, 2018. https://europe.edf.org/news/2018/31/07/alternative-fuels-future-zero-emissions-shipping

Burbridge, D. Building a lower-carbon construction industry. *Carbon Trust*, February 1, 2012. https://www.carbontrust.com/news/2012/02/building-a-lower-carbon-construction-industry-1/

Burger, A. First U.S. grid-connected OTEC plant goes live on Hawaii. *Renewable Energy World*, September 1, 2015. https://www.renewableenergyworld.com/2015/09/01/first-u-s-grid-connected-otec-plant-goes-live-on-hawaii/#gref

Canadell, P., and Jackson, R. Turning methane into carbon dioxide could help us fight climate change. *The Conversation*, May 21, 2019. https://theconversation.com/turning-methane-into-carbon-dioxide-could-help-us-fight-climate-change-117317

CEMBUREAU The European Cement Association. Building carbon neutrality in Europe: engaging for concrete solution. CEMBUREAU, D/2018/5457/October, October 2018. https://lowcarboneconomy.cembureau.eu/wp-content/uploads/2018/12/CEMBUREAU-BUILDING-CARBON-NEUTRALITY-IN-EUROPE_final.pdf

Coady, D., et al. Global fossil fuel subsidies remain large: an update based on country-level estimates. International Monetary Fund working paper WP/19/89, May 2019. https://www.imf.org/en/Publications/WP/Issues/2019/05/02/Global-Fossil-Fuel-Subsidies-Remain-Large-An-Update-Based-on-Country-Level-Estimates-46509

Conniff, R. The greenhouse gas that nobody knew. *Yale Environment 360*, November 13, 2008. https://e360.yale.edu/features/the_greenhouse_gasthat_nobody_knew

Construction Climate Challenge. *Green building materials that are alternatives to concrete*. July 7, 2016. https://constructionclimatechallenge.com/2016/07/07/green-building-materials-that-are-alternatives-to-concrete/

Coppola, L., et al. Binders alternative to Portland cement and waste management for sustainable construction—part 2. *Journal of Applied Biomaterials & Functional Materials*, 16, 207–221, 2018. https://journals.sagepub.com/doi/pdf/10.1177/2280800018782852

Cronin, J., et al. Climate change impacts on the energy system: a review of trends and gaps. *Climatic Change*, 151, 79–93, 2018. https://link.springer.com/article/10.1007/s10584-018-2265-4

Davidson, S. Sustainable bioenergy: genomics and biofuels development. *Nature Education*, 1, 175, 2008. https://www.nature.com/scitable/topicpage/sustainable-bioenergy-genomics-and-biofuels-development-44571/#

Deemer, B.R., et al. Greenhouse gas emissions from reservoir water surfaces: a new global synthesis. *Bioscience*, 66, 949–964, 2016. https://academic.oup.com/bioscience/article/66/11/949/2754271

Deign, J. UK wave energy startups, clamoring for government money, have failed to deliver. *GreenTechMedia*, November 20, 2017. https://www.greentechmedia.com/articles/read/uk-wave-energy-startups-have-failed-on-their-promises#gs.l6xnhn

Denholm, P., et al. Land-use requirements of modern wind power plants in the United States. National Renewable Energy Laboratory Technical Report Technical Report NREL/TP-6A2-45834, August 2009. https://www.nrel.gov/docs/fy09osti/45834.pdf

Ding, Y., et al. Biochar to improve soil fertility. A review. *Agronomy for Sustainable Development*, 36, 36, 2016. https://link.springer.com/article/10.1007/s13593-016-0372-z

DNV-GL Maritime. *Assessment of selected alternative fuels and technologies*. April 2018. https://hydrogeneurope.eu/sites/default/files/2018-04/DNV_GL_Complete-Alt-Fuels_guidance_paper_2018-04_web.pdf

Dunuweera, S.P., and Rajapakse, R.M.G. Cement types, composition, uses and advantages of nanocement, environmental impact on cement production, and possible solutions. *Advances in Materials Science and Engineering*, 2018, Article ID 4158682, doi: 10.1155/2018/4158682, 2018. http://downloads.hindawi.com/journals/amse/2018/4158682.pdf

DutchNews.nl. *Motorway speeds reduced because of nitrogen reduction failings*. September 10, 2019. https://www.dutchnews.nl/news/2019/09/motorway-speeds-reduced-because-of-nitrogen-reduction-failings/

En:former. *How hydrogen could change the face of steel production as we know it.* May 29, 2019. https://www.en-former.com/en/hydrogen-revolution-steel-production/

European Environment Agency. Emissions and supply of fluorinated greenhouse gases. Indicator Assessment Prod-ID: IND-354-en, September 9, 2019. https://www.eea.europa.eu/data-and-maps/indicators/emissions-and-consumption-of-fluorinated-2/assessment-1

European Technology and Innovation Platform. Biofuels for rail transport. ETIP-B-SABS 2, 2019. http://www.etipbioenergy.eu/?option=com_content&view=article&id=295.

Faaij, A. Biomass resources, worldwide. In: Meyers, R.A. (ed.). *Encyclopedia of Sustainability Science and Technology*, pp. 1531–1583, Springer, New York, 2012. https://link.springer.com/content/pdf/10.1007%2F978-1-4419-0851-3_386.pdf

Fischedick, M., et al. Techno-economic evaluation of innovative steel production technologies. *Journal of Cleaner Production, 84*, 563–580, 2014. https://www.sciencedirect.com/science/article/abs/pii/S095965261400540X

Flynn, K.J. Algal biofuel production is neither environmentally nor commercially sustainable. *The Conversation*, August 8, 2017. https://theconversation.com/algal-biofuel-production-is-neither-environmentally-nor-commercially-sustainable-82095

Frank, S., et al. Reducing greenhouse gas emissions in agriculture without compromising food security? *Environmental Research Letters, 12*, 105004, 2017. https://iopscience.iop.org/article/10.1088/1748-9326/aa8c83/pdf

Futurefeed. *Feeding livestock a seaweed supplement called FutureFeed could simultaneously help to secure global food security and fight climate change by reducing powerful greenhouse gas emissions.* May 31, 2019. https://www.csiro.au/en/Research/AF/Areas/Food-security/FutureFeed

Gilbert, P., et al. Assessment of full life-cycle air emissions of alternative shipping fuels. *Journal of Cleaner Production, 172*, 855–866, 2018. https://www.sciencedirect.com/science/article/pii/S0959652617324721?via%3Dihub

Gillingham, K. Carbon calculus. *Finance and Development, 56*, December 2019. https://www.imf.org/external/pubs/ft/fandd/2019/12/the-true-cost-of-reducing-greenhouse-gas-emissions-gillingham.htm

Glaser, B., et al. The Terra Preta phenomenon: a model for sustainable agriculture in the humid tropics. *Naturwissenschaften, 88*, 37–41, 2001. https://link.springer.com/article/10.1007/s001140000193

Global Opportunity Explorer. *Using $CO_2$ to make concrete.* November 9, 2018. https://goexplorer.org/using-co2-to-make-concrete/

Goldstein, B., et al. Geothermal Energy. *IPCC Special Report on Renewable Energy Sources and Climate Change Mitigation*. Cambridge University Press, 436 pp., Cambridge, UK, and New York, 2011. http://www.ipcc-wg3.de/report/IPCC_SRREN_Ch04.pdf

Goodwin, P.A., et al. Pathways to 1.5°C and 2°C warming based on observational and geological constraints. *Nature Geoscience*, *11*, 102107, 2018. https://www.nature.com/articles/s41561-017-0054-8

Grandelli, P., et al. Modeling the physical and biochemical influence of ocean thermal energy conversion plant discharges into their adjacent waters. US Department of Energy–Office of Scientific and Technical Information, 2012. https://www.osti.gov/servlets/purl/1055480

Gretz, J., et al. Hydrogen in the steel industry. *International Journal for Hydrogen Energy*, *16*, 691–693, 1991. https://www.sciencedirect.com/science/article/pii/036031999190193M

Harding, R. Japan's hydrogen dream: game-changer or a lot of hot air? *Financial Times Special Report on How we will live in 2050*, June 17, 2019. https://www.ft.com/content/c586475e-7260-11e9-bf5c-6eeb837566c5?sfns=mo

Harrabin, R. Plant-based diet can fight climate change—UN. *BBC News*, August 8, 2019. https://www.bbc.com/news/science-environment-49238749

Hayes, C.J., et al. Chapter three—combustion pathways of biofuel model compounds: a review of recent research and current challenges pertaining to first-, second-, and third-generation biofuels. *Advances in Physical Organic Chemistry*, *49*, 103–187, 2015. https://www.sciencedirect.com/science/article/pii/S0065316015000052

Heimann, K. Explainer: what are algal biofuels. *The Conversation*, April 9, 2013. https://theconversation.com/explainer-what-are-algal-biofuels-12560

HELIOSCSP. Dubai 950 MW Concentrated Solar Power +PV Noor Energy 1 breaks 8 world records. *Solar Thermal Energy News*, April 16, 2019. http://helioscsp.com/dubai-950-mw-concentrated-solar-power-pv-noor-energy-1-breaks-8-world-records/

Hesse, H.C., et al. Lithium-ion battery storage for the grid—a review of stationary battery storage system design tailored for applications in modern power grids. *Energies*, *10*, 2107, 2017. https://www.mdpi.com/1996-1073/10/12/2107

Iceland Magazine. Does Iceland really produce all of its electricity from renewables? January 15, 2019. https://icelandmag.is/article/does-iceland-really-produce-all-its-electricity-renewables

International Atomic Energy Association. Hydrogen production using nuclear energy. IAEA nuclear ENERGY series, NP-T-4.2. 2013. https://www-pub.iaea.org/MTCD/Publications/PDF/Pub1577_web.pdf

International Energy Agency. *Global energy and CO$_2$ status report: the latest trends in energy and emissions in 2018*. 2019. https://www.iea.org/geco/emissions/

International Energy Agency. *Transport biofuels*. May 27, 2019. https://www.iea.org/tcep/transport/biofuels/

International Hydropower Association. *Types of hydropower*. 2020. https://www.hydropower.org/types-of-hydropower

International Renewable Energy Agency (IRENA). *Geothermal power: technology brief*. Abu Dhabi, 2017. https://www.irena.org/-/media/Files/IRENA/Agency/Publication/2017/Aug/IRENA_Geothermal_Power_2017.pdf

International Renewable Energy Agency (IRENA). *Global energy transformation: a roadmap to 2050*. Abu Dhabi, 2018. https://www.irena.org/-/media/Files/IRENA/Agency/Publication/2018/Apr/IRENA_Report_GET_2018.pdf

International Renewable Energy Agency (IRENA). *Geothermal energy*, Abu Dhabi, 2018. https://www.irena.org/geothermal

International Renewable Energy Agency (IRENA). *Renewable power generation costs in 2018*. Abu Dhabi, 2019. https://www.irena.org/-/media/Files/IRENA/Agency/Publication/2019/May/IRENA_Renewable-Power-Generations-Costs-in-2018.pdf?la=en&hash=99683CDDBC40A729A5F51C20DA7B6C297F794C5D

IPCC. *Climate change and land: an IPCC special report on climate change, desertification, land degradation, sustainable land management, food security, and greenhouse gas fluxes in terrestrial ecosystems*. Shukla, P.R. et al. (eds.), 2019. https://www.ipcc.ch/site/assets/uploads/2019/11/SRCCL-Full-Report-Compiled-191128.pdf

Irfan, U. Bitcoin is an energy hog. Where is all that electricity coming from? A new report claims it's mostly powered by renewables. Be skeptical. *Vox*, June 18, 2019. https://www.vox.com/2019/6/18/18642645/bitcoin-energy-price-renewable-china

Jaczko, G. I oversaw the U.S. nuclear power industry. Now I think it should be banned. *Washington Post*, May 17, 2019. https://www.washingtonpost.com/outlook/i-oversaw-the-us-nuclear-power-industry-now-i-think-it-should-be-banned/2019/05/16/a3b8be52-71db-11e9-9eb4-0828f5389013_story.html?noredirect=on

Karell, M., and Chattopadhyay, A. NOx emission reduction strategies. *Pollution Online*, June 16, 2000. https://www.pollutiononline.com/doc/nox-emission-reduction-strategies-0001

Kellner, T. A Towering achievement: this summer in Holland, GE will build the world's largest wind turbine. GE Reports, January 18, 2019. https://www.ge.com/reports/towering-achievement-summer-holland-ge-will-build-worlds-largest-wind-turbine/

Kenny, P., and Flynn, K.J. Physiology limits commercially viable photoautotrophic production of microalgal biofuels. *Journal of Applied Phycology*, 29, 2713–2727, 2017. https://link.springer.com/article/10.1007/s10811-017-1214-3

Kittredge, J. Soil carbon restoration: can biology do the job? Northeast Organic Farming Association/Massachusetts Chapter, Inc., August 14, 2015. http://www. nofamass.org/sites/default/files/2015_White_Paper_web.pdf

Knitterscheidt, K. ThyssenKrupp steels itself for a carbon-free future. *Handelsblatt Today*, January 23, 2019. https://www.handelsblatt.com/today/companies/ green-energy-thyssenkrupp-steels-itself-for-a-carbon-free-future/23894808. html?ticket=ST-5091278-0kupmO20bNQEddkJb7Dq-ap6

Kołwzan, K., and Narewski, M. Alternative fuels for marine applications. *Latvian Journal of Chemistry*, 4, 398–406, 2012. https://www.researchgate.net/publication/ 264972038_Alternative_Fuels_for_Marine_Applications

Kraemer, S. Missing link for solar hydrogen is ... ammonia? *PhysOrg*, January 9, 2018. https://phys.org/news/2018-01-link-solar-hydrogen-ammonia.html

Kristmannsdóttir, H., and Ármannsson, H. Environmental aspects of geothermal energy utilization. *Geothermics*, 32, 451–461, 2003. https://www.sciencedirect.com/ science/article/pii/S037565050300052X

Leister, D. Genetic engineering, synthetic biology and the light reactions of photosynthesis. *Plant Physiology*, 179, 778–793, 2019. http://www.plantphysiol.org/content/plantphysiol/179/3/778.full.pdf

Levy, N. Snøhetta unveils plans for "energy-positive" Arctic Circle hotel. *DeZeen*, February 12, 2018. https://www.dezeen.com/2018/02/12/snohetta-energy-positive-sustainable-hotel-arctic-circle-norway/

Liu, X., et al. Modular engineering for efficient photosynthetic biosynthesis of 1-butanol from $CO_2$ in cyanobacteria. *Energy and Environmental Science*, 2019. doi: 10.1039/c9ee01214a. https://pubs.rsc.org/en/content/articlelanding/2019/ee/ c9ee01214a#!divAbstract

Luckow, P., et al. Technical and institutional barriers to the expansion of wind and solar energy: near-term measures to foster development. Synapse Energy Economics Inc., June 19, 2015. https://synapse-energy.com/sites/default/files/ Barriers-to-Wind-and-Solar-15-047_0.pdf

Lund, J.W. The USA geothermal country update. *Geothermics*, 32, 409–418, 2003. https://www.sciencedirect.com/science/article/pii/S0375650503000531?via%3Dihub

Lyons Hardcastle, J. New building material twice as strong as concrete, low carbon footprint. *Environmental Leader*, August 5, 2015. https://www. environmentalleader.com/2015/08/new-building-material-twice-as-strong-as-concrete-low-carbon-footprint/

Madaeni, S.H., et al. Capacity value of concentrating solar power plants. National Renewable Energy Laboratory Technical Report NREL/TP-6A20-51253, June 2011. https://www.nrel.gov/docs/fy11osti/51253.pdf

Mahajan, M. Plunging prices mean building new renewable energy is cheaper than running existing coal. *Forbes Energy Innovation: Policy and Technology*, December 3, 2018. https://www.forbes.com/sites/energyinnovation/2018/12/03/plunging-prices-mean-building-new-renewable-energy-is-cheaper-than-running-existing-coal/#4bd7930731f3

*Marine and Hydrokinetic Technology Database*, January 9, 2018. https://openei.org/wiki/Marine_and_Hydrokinetic_Technology_Database

Martins Adeniyi, O., et al. Algae biofuel current status and future applications. *Renewable and Sustainable Energy Reviews*, 90, 316–335, 2018. https://www.sciencedirect.com/science/article/pii/S1364032118301552

Matassa, S., et al. Can direct conversion of used nitrogen to new feed and protein help feed the world? *Environmental Science and Technology*, 49, 5247–5254, 2015. https://www.powertoprotein.eu/wp-content/uploads/can_direct_conversion_of_used_nitrogen_to_new_feed_and_protein_help_feed_the_world_1.pdf

Matassa, S., et al. Microbial protein: future sustainable food supply route with low environmental footprint. *Microbial Biotechnology*, 9, 568–575, 2016. https://www.ncbi.nlm.nih.gov/pmc/articles/PMC4993174/

Maxwell, S.L., et al. Degradation and forgone removals increase the carbon impact of intact forest loss by 626%. *Science Advances*, 5, eaax2546, 2019. https://advances.sciencemag.org/content/5/10/eaax2546.full

Millar, N., et al. Management of nitrogen fertilizer to reduce nitrous oxide ($N_2O$) emissions from field crops. Michigan State University Extension Bulletin E3152, November 2014. https://www.canr.msu.edu/uploads/resources/pdfs/management_of_nitrogen_fertiler_(e3152).pdf

Minter, A. China's hydrogen economy is coming. *Bloomberg Opinion*, March 23, 2019. https://www.bloomberg.com/opinion/articles/2019-03-23/now-china-wants-to-lead-the-world-in-hydrogen-fuel-cells

MIT Technology Review. *Germany runs up against the limits of renewables.* May 24, 2016. https://www.technologyreview.com/s/601514/germany-runs-up-against-the-limits-of-renewables/

Mittelbach, M. Biodiesel. In: Meyers, R.A. (ed.). *Encyclopedia of sustainability science and technology*, pp. 921–938, Springer, New York, 2012. https://link.springer.com/content/pdf/10.1007%2F978-1-4419-0851-3_386.pdf

Muller-Langer, S., et al. Biofuels: a technical, economic and environmental comparison. In: Meyers, R.A. (ed.). *Encyclopedia of sustainability science and technology*, pp. 1039–1066, Springer, New York, 2012. https://link.springer.com/content/pdf/10.1007%2F978-1-4419-0851-3_386.pdf

National Academies of Sciences, Engineering, and Medicine. *Commercial aircraft propulsion and energy systems research: reducing global carbon emissions.* The National Academies Press, Washington, DC, 2016. doi:10.17226/23490. https:// www.nap.edu/read/23490/chapter/1

Nisbet, E.G., et al. Methane mitigation: methods to reduce emissions, on the path to the Paris Agreement. *Reviews of Geophysics, 58,* e2019RG000675, 2020. https:// agupubs.onlinelibrary.wiley.com/doi/epdf/10.1029/2019RG000675

Norwegian Ministry of Petroleum and Energy. *Renewable energy production in Norway.* https://www.regjeringen.no/en/topics/energy/renewable-energy/ renewable-energy-production-in-norway/id2343462/

Overheu, T., and Garlinge, J. Regenerative agriculture and pastoralism in Western Australia. Government of Western Australia—Department of Primary Industries and Regional Development, August 20, 2019. https://www.agric.wa.gov.au/land-use/regenerative-agriculture-and-pastoralism-western-australia

Øverland, M. *An animal feed revolution: how microbial protein sources will create a sustainable future.* Norwegian University of Life Sciences, NMBU, presentation. https://www.feedproteinvision.com/wp-content/uploads/2018/03/Day-2-Margareth-Overland.pdf

Owaid, H.M., et al. A review of sustainable supplementary cementitious materials as an alternative to all-Portland cement mortar and concrete. *Australian Journal of Basic and Applied Sciences, 6,* 287–303, 2012. https://pdfs.semanticscholar.org/ b304/ae07b02ba1a9edd33fc20080871a895519f6.pdf?_ga=2.98691166.196567878 3.1564094996-23631970.1563720700

Ożadowicz, A., and Grela, J. Energy saving in the street lighting control system—a new approach based on the EN-15232 standard. *Energy Efficiency, 10,* 563–576, 2017. https://link.springer.com/article/10.1007/s12053-016-9476-1

Paustian, K., et al. Chapter 1, Introduction: Agriculture, Forestry and Other Land Use (AFOLU). In: Eggleston, S., et al. (eds.). *IPCC guidelines for national greenhouse gas inventories, Volume 4: Agriculture, forestry and other land use,* 21 pp., IGES, Japan, 2006. https://www.ipcc-nggip.iges.or.jp/public/2006gl/pdf/4_Volume4/ V4_01_Ch1_Introduction.pdf

Petley, D. The Vaiont (Vajont) landslide of 1963. *AGU The Landslide Blog,* December 11, 2008. https://blogs.agu.org/landslideblog/2008/12/11/the-vaiont-vajont-landslide-of-1963/

Pikaar, I., et al. Carbon emission avoidance and capture by producing in-reactor microbial biomass based food, feed and slow release fertilizer: potentials and limitations. *Science of the Total Environment, 644,* 1525–1530, 2018. https://www. sciencedirect.com/science/article/pii/S0048969718325749?via%3Dihub

Qidwai, A., et al. Introduction of nanotechnology in the field of biofuel production. In: Srivastava, N., et al. (eds.). *Green nanotechnology for biofuel production, biofuel and biorefinery technologies 5*, pp 29–38, Springer International Publishing AG, Switzerland, 2018. https://www.researchgate.net/publication/323849526_ Introduction_of_Nanotechnology_in_the_Field_of_Biofuel_Production

Räsänen, T. et al. Greenhouse gas emissions of hydropower in the Mekong River Basin. *Environmental Research Letters, 13*, 034030, 2018. https://iopscience.iop. org/article/10.1088/1748-9326/aaa817/pdf

*Regeneration International*. 2019. https://regenerationinternational.org/why-regenerative- agriculture/

REN21. *Renewables 2018 global status report*. REN21 Secretariat, Paris, ISBN 978- 3-9818911-3-3, 2018. http://www.ren21.net/wp-content/uploads/2018/06/17- 8652_GSR2018_FullReport_web_final_.pdf

Renewable Resources Coalition. *11 different sources of alternative energy*. 2016. https://www.renewableresourcescoalition.org/alternative-energy-sources/

RGN, Resource Global Network. GE to build $400 million Haliade-X—the world's biggest wind turbine. *RGN News*, March 5, 2018. https://resourceglobalnetwork.com/ 2018/03/05/ge-to-build-400-million-haliade-x-the-worlds-biggest-wind-turbine/

Rissman, J. Concrete change: making cement carbon-negative. *Greenbiz*, December 6, 2018. https://www.greenbiz.com/article/concrete-change-making- cement-carbon-negative

Rocheleau, G.J., and Grandelli, P. Physical and biological modeling of a 100 megawatt Ocean Thermal Energy Conversion discharge plume. Oceans'11 MTS/IEEE Kona, pp. 1–10, 2011. https://ieeexplore.ieee.org/document/6107077

Rodgers, L. Climate change: the massive $CO_2$ emitter you may not know about. *BBC News—Science and Environment*, December 17, 2018. https://www.bbc.com/ news/science-environment-46455844

Romm, J. Climate and hydrogen car advocate gets almost everything wrong about plug-in cars. *Climate Progress*, October 6, 2009. http://climateprogress.org//2009/ 10/06/climate-and-hydrogen-car-advocate-gets-almost-everything-wrong- about-plug-in-cars/

Shackley, S., et al. Biochar, tool for climate change mitigation and soil management. In: Meyers, R.A. (ed.). *Encyclopedia of sustainability science and technology*, pp. 845–893, Springer, New York, 2012. https://link.springer.com/content/pdf/ 10.1007%2F978-1-4419-0851-3_386.pdf

Sigfússon, B. and Uihlein, A. Technology, market and economic aspects of geothermal energy in Europe. JRC Geothermal Energy Status Report; EUR 27623 EN,

2015. https://setis.ec.europa.eu/sites/default/files/reports/2015_jrc_geothermal_energy_status_report.pdf

Skinner, I., et al. Railways and biofuel. First UIC Report, July, 2007. https://uic.org/IMG/pdf/railways_and_biofuels_final_report.pdf

Solar Wind Energy Inc. *Solar wind energy tower.* Last accessed 23 June 2019. https://www.solarwindenergytower.com/the-tower.html#towervid

Solaun, K., and Cerdá, E. Climate change impacts on renewable energy generation. A review of quantitative projections. *Renewable and Sustainable Energy Reviews, 116,* 109415, 2019. https://www.sciencedirect.com/science/article/pii/S1364032119306239

Stangarona, T. South Korea's hydrogen economy ambitions. *The Diplomat,* January 31, 2019. https://thediplomat.com/2019/01/south-koreas-hydrogen-economy-ambitions/

Stephen, S. Efforts afoot to tap electric car batteries to power homes. *The Japan Times,* April 17, 2018. https://www.japantimes.co.jp/news/2018/04/17/business/efforts-afoot-tap-electric-car-batteries-power-homes/#.XQ8xkK2B1qw

Sturges, W.T., et al. Emissions halted of the potent greenhouse gas $SF_5CF_3$. *Atmospheric Chemistry and Physics, 12,* 3653–3658, 2012. https://hal.archives-ouvertes.fr/hal-00821199/documentTanzer, S.E., and Ramirez, A. When are negative emissions negative emissions? *Energy and Environmental Science, 12,* 1210, 2019. https://pubs.rsc.org/en/content/articlelanding/2019/ee/c8ee03338b#!divAbstract

The Maritime Executive. New research shows benefits of ammonia as marine fuel. June 11, 2019. https://www.maritime-executive.com/article/new-research-shows-benefits-of-ammonia-as-marine-fuel

Tidball, R., et al. Cost and performance assumptions for modeling electricity generation technologies. US National Renewable Energy Laboratory, November 2010, 211 pp., 2010. https://www.nrel.gov/docs/fy11osti/48595.pdf

Total Materia. Hydrogen in steels. Total Materia Article, August 2007. https://www.totalmateria.com/page.aspx?ID=CheckArticle&site=kts&NM=206

United States Environment Protection Agency. Overview of $SF_6$ emissions sources and reduction options in electric power systems. EPA Climate Change Division (6207J), EPA 430-R-18-004, August 2018. https://www.epa.gov/sites/production/files/2018-08/documents/12183_sf6_partnership_overview_v20_release_508.pdf

United States Environmental Protection Agency, Office of Air and Radiation. F-GHG emissions reduction efforts: flat panel display supplier profiles. May 2013. https://www.epa.gov/sites/production/files/2015-07/documents/supplier_profiles_2013.pdf

van de Schoot, E. Windmolen lekt extreem schadelijk gas. *De Telegraaf Financieel Nieuws*, October 29, 2019. https://www.telegraaf.nl/financieel/921218615/windmolen-lekt-extreem-schadelijk-gas

Venkatarama Reddy, B.V. Sustainable materials for low carbon buildings. *International Journal of Low-Carbon Technologies*, 4, 175–181, 2009. https://academic.oup.com/ijlct/article/4/3/175/710965

Wallmann, K., and Aloisi, G. The global carbon cycle: geological processes. In: Knoll, A.H., Canfield, D.E., and Konhauser, K.O. (eds.). *Fundamentals of geobiology*, First Edition, pp. 20–35, Blackwell Publishing Ltd., Chichester, UK, 2012. http://geosci.uchicago.edu/~kite/doc/Fundamentals_of_Geobiology_Chapter_3.pdf

Watson, D. Bioclimatic design. In: Meyers, R.A. (ed.). *Encyclopedia of sustainability science and technology*, pp. 893–920, Springer, New York, 2012. https://link.springer.com/content/pdf/10.1007%2F978-1-4419-0851-3_386.pdf

Wikipedia. *Geothermal heating*. Last accessed June 28, 2019. https://en.wikipedia.org/wiki/Geothermal_heating

Willemsen, N. The rapid development of geothermal energy in the Netherlands. *Think GeoEnergy—Geothermal Energy News*, September 8, 2016. http://www.thinkgeoenergy.com/the-rapid-development-of-geothermal-energy-in-the-netherlands/

Winkless, L. Dutch trains are now powered by wind. *Forbes*, January 12, 2017. https://www.forbes.com/sites/lauriewinkless/2017/01/12/dutch-trains-are-now-powered-by-wind/#1d6707d12d29

Wood, J. Welcome to a hydrogen-powered world. *Mitsubishi Heavy Industries Group—Spectra*, January 8, 2019. https://spectra.mhi.com/welcome-to-a-hydrogen-powered-world

World Economic Forum. *Which countries produce the most hydroelectric power?* October 13, 2015. https://www.weforum.org/agenda/2015/10/which-countries-produce-the-most-hydroelectric-power/

World Nuclear Association. *Transport and the hydrogen economy*. 2018. http://www.world-nuclear.org/information-library/non-power-nuclear-applications/transport/transport-and-the-hydrogen-economy.aspx

World Steel Association. Steel's contribution to a low carbon future and climate resilient societies. Worldsteel Position Paper, 2019. https://www.worldsteel.org/en/dam/jcr:7ec64bc1-c51c-439b-84b8-94496686b8c6/Position_paper_climate_2019_vfinal.pdf

World Wind Energy Association. *Wind power capacity worldwide reaches 597 GW, 50.1 GW added in 2018*. February 2019. https://wwindea.org/blog/2019/02/25/wind-power-capacity-worldwide-reaches-600-gw-539-gw-added-in-2018/

Würsig, G.M. Alternative fuels: the options. *DNV-GL Maritime Impact*, October 9, 2018. https://www.dnvgl.com/expert-story/maritime-impact/alternative-fuels. html

Yu, S., and Jain, P.K. Plasmonic photosynthesis of C1–C3 hydrocarbons from carbon dioxide assisted by an ionic liquid. *Nature Communications*, *10*, 2019. https://www. nature.com/articles/s41467-019-10084-5

Zhang, D., and Song, J. Mechanisms for geological carbon sequestration. *Procedia IUTAM*, *10*, 219–327, 2014. https://www.sciencedirect.com/science/article/pii/ S2210983814000285

# 4

# The New Kid on the Block

Negative Emissions through Greenhouse
Gas Removal

## 4.1. Setting the Scene: The Gigatons Challenge

Over the years, we have come to realize that that there are acceptable limits to the atmospheric concentrations of the human-caused greenhouse gases that have accumulated in the atmosphere. This does not mean that those limits are desirable; it merely means that they are generally considered acceptable enough to be lived with. Incidentally, the same applies to other human-caused emissions. For toxic pollutants such as lead or radioactivity, for example, internationally agreed acceptable levels or concentrations are set with respect to their health implications. For greenhouse gases, international agreements focus less on their exact concentrations in the atmosphere and more on the total climatic impact of the gases. In that respect, the 2015 Paris Agreement states that greenhouse gas concentrations ought to be limited so that global mean warming remains below 2°C at most, and 1.5°C by preference, relative to global mean temperatures of about 150 to 200 years ago, before the industrial revolution. If we exceed the greenhouse gas levels that correspond to those maximum warming targets, then we need to both curtail further emissions (Chapter 3) and remove any excess of accumulated gases (this chapter). Net removal of greenhouse gases has become known as negative emissions.

It is not every day that we hear that something called negative can in fact be positive. Yet, this is exactly the case with the term negative emissions. It doesn't mean dangerous emissions or detrimental emissions. It simply means the opposite of positive emissions. Mathematically speaking, positive emissions are the input of gases into the climate system as a result of humanity's various activities, while negative emissions concern the removal, or drawdown, of such gases from the atmosphere into safe storage systems from which they don't return. There is nothing magical or complicated about this in concept. But the term negative emissions is still greatly misunderstood and misused in practice. Tanzer and Ramirez (2019) endeavored to clearly define the issue, so that methods can no longer be branded negative emissions unless they adhere to strict conditions. They state: "an overarching necessity is to ensure that the total effect of all components within the complex system of a negative emissions technology (NET) is the permanent removal of atmospheric greenhouse gases, and thereby a net decrease in the greenhouse gas concentration in the atmosphere."

For practical implementations, Tanzer and Ramirez (2019) break the overarching requirement down into delivery on four components: (1) removal of greenhouse gases from the atmosphere, (2) long-term or permanent storage of the removed greenhouse gases, (3) full accounting for emissions associated with both the upstream and downstream supply chains of the negative emission technology (that is, full life-cycle emissions assessment), and (4) contributions to storage of non-atmospheric greenhouse gases, re-emission of captured gases to the atmosphere, and identification and accounting for emissions that are avoided emissions rather than negative emissions. While this may seem onerous, it's essential to ensure that methods are only branded NETs when they really are NETs, which places critical constraints on the way nations and industries can present their carbon accounts and budgets.

NETs are less maturely developed than emissions reduction methodologies. Most of the world has only woken up to the need for NETs since 2013 or so. Fortunately, a handful of pioneering researchers realized much earlier that NETs might become essential and made a start decades ago. Many of the current concepts and plans still draw to some extent upon the work of those pioneers, or on derivatives of their initial work. Some of the pioneers used geochemical knowhow to propose NETs based on artificial acceleration of the weathering of silicate minerals, notably olivine. Such weathering is a key process for $CO_2$ control in natural, geological cycles, albeit on very slow timescales. Other pioneers initiated more technological approaches, which have branched into a wide array of techniques focused on development of chemicals, membranes, polymers, and other media that might eventually be scaled up into industrial facilities for direct capture of $CO_2$ from air. Still others began looking deeply at restoring soil-carbon levels around the world; at carbon drawdown through global reforestation; at feeding lime or ground silicate rock dust into the ocean to both make the ocean absorb more $CO_2$ and counteract ocean acidification; and at fertilizing the ocean to trigger algal blooms that consume $CO_2$ and then sink into the deep sea.

Over time, more and more new ideas or variants of ideas have emerged. Critically, however, none has so far been tested at scales of millions (or more) tons of $CO_2$, even though we will need them at billion-ton scales within several decades. As we saw in Chapter 2, we will need NETs to achieve 70–280 GtC (that is, 260–1030 $GtCO_2$) of total removal by 2100. Properly scaled and implemented, NETs—alongside rapid emissions reduction and avoidance programs—can still help us bring climate change within acceptable levels by the end of the century, and keep it there. But this requires rapid and focused engagement in a wide portfolio of NETs research, development, testing, and upscaling.

Here, then, we encounter what I like to call the *Gigatons challenge*. It is the challenge of taking all reasonable ideas, concepts, and small-scale trials and scaling them up rapidly and massively to demonstrate real-life potential, economic and environmental feasibility, and social acceptability. No stone can be left unturned. The required amount of carbon removal is so large that no single method will ever suffice on its own; we will need a broad portfolio of methods working in parallel. And no shortcuts can be allowed in dispassionate emissions assessments of all methods across their entire life cycle of processes. The next requirement in the Gigatons challenge is that any successful method has to be scalable to million-ton or billion-ton scales of carbon removal. It also has to be environmentally sound, in that it should be feasible without jeopardizing biodiversity or global food production, and it has to be economically feasible and socially acceptable. Finally, to ensure equitable and safe implementation, it needs to be couched within appropriate international environmental legal and governance contexts.

Given these requirements, the Gigatons NETs challenge cuts across many disciplines, including at least Earth Science, Biology, Engineering, Chemistry, Physics, Economics, Environmental Law, Politics, and Social Science. Interactions across these disciplines are arranged through problem-oriented collaborations that often also include industry, development of overarching institutes, new journals with wide-ranging disciplinary remits, and special meetings and conferences. It is evident that the sheer quantity of preparatory work and organization needed before large-scale implementation becomes feasible requires that work should commence imminently. In consequence, developing appropriately scaled NETs calls for high levels of strategic investment. In return, it promises to open up vast amounts of new job opportunities, technologies, manufacturing, sustainable investment opportunities, and so on.

And then a word of caution: a common mistake with determining how effective a method could be for reducing atmospheric $CO_2$ levels is that the return of $CO_2$ from the oceans is overlooked. As we emitted $CO_2$, $CO_2$ pressures in the atmosphere went up, and air-sea gas exchange responded to that change in the air-sea gradient. As a result, roughly a third of the emissions got absorbed by the oceans, where it resulted in ocean acidification. If and when we manage to reduce the atmospheric $CO_2$ level, this chain of events starts to operate in the opposite direction; the oceans will then outgas $CO_2$ into the atmosphere. As a result, 1.6–1.7 times more $CO_2$ must be captured from the atmosphere to achieve a certain $CO_2$ reduction than we would think on the basis of considering the atmosphere alone.

Let's look at an example. A 1 ppm $CO_2$ concentration change in the atmosphere alone stands for a mass change of about 2.12 GtC, or 7.81 $GtCO_2$. But, because of the oceans, reducing the atmospheric $CO_2$ concentration by 1 ppm will in fact require the removal of roughly 3.5 GtC, or 12.9 $GtCO_2$. This is easily overlooked, yet fundamental when we start talking about trapping $CO_2$ directly from the atmosphere. When I reported amounts of carbon removal needed in Chapter 2, this effect was accounted for. But when reports look at efficacies of $CO_2$ removal systems, the effect is not commonly taken into account when comparing with atmospheric carbon inventories. Be sure to keep the ocean effect in mind when reading through the literature. To avoid confusion, it is best to compare reported carbon removal efficacies with carefully calculated net carbon removal requirements from the entire system (e.g., Rogelj et al., 2018).

And then we need to raise another word of caution. NETs, and more specifically carbon dioxide removal (CDR), cannot be seen as an easy fix to be applied without emissions reduction and avoidance of adding new emissions. This is because $CO_2$ emissions cause not only climate change but also ocean acidification, which is a consequence of oceanic absorption of about a third of the emissions. Our current, so-called business-as-usual, pathway of emissions that now amount to more than 10 GtC per year is causing the rapid build-up of extensive ocean acidification. Alarmingly, it turns out that if we keep emitting as we are and apply NETs for a CDR component of 5 GtC per year (which means we'd be doing really well), we would hardly improve the ocean acidification situation. In fact, we'd still only make a modest improvement to ocean acidification even for a CDR component of 25 GtC per year, which is beyond realistic imagination. Worse, these improvements are measured after 700 years; on shorter timescales, the improvements are even smaller! This is the case because ocean acidification is a process that would re-adjust on deep ocean mixing timescales, which span many centuries to a millennium. From an ocean acidification point of view—a critical one as far as marine ecosystems are concerned—CDR is therefore definitely not a viable alternative to drastic emissions reduction and avoidance. Those two have to be implemented immediately, and CDR can then be applied in parallel. For details on this, see Mathesius et al. (2015).

With these points in mind, let's get back to the gist of this chapter. NETs can be grouped in various ways. I'm going to go with my own favorite subdivision. It distinguishes between fundamentally Earth System–based NETs and technology-based NETs, with these main categories subdivided into land-based and marine NETs. Accordingly, section 4.2 discusses land-based Earth System NETs, section 4.3 marine Earth System NETs, section 4.4 land-based technological NETs, and section 4.5 marine technological NETs. In section 4.6, I synthesize the information by setting up a simple index that amalgamates scores for the carbon removal potential, environmental impacts, and technological readiness of the different approaches, supported by a table with relevant input values. I do not discuss costs too much, except when they are used to illustrate specific issues, but briefly get back to costs later, in Chapter 8. Again, key references for specific points are given in the text, and—to keep the text accessible—most supporting literature is listed at the end of the chapter without specific call-outs in the text.

## 4.2. Land-based Earth System NETs

In this group, there are three main streams. The first concerns carbon drawdown from the climate system by growing vegetation and increasing soil carbon concentrations. Often, this involves remedial activities where forest expansion is undertaken to compensate for loss that has resulted from a long history of deforestation, or where soil carbon contents are restored back to what they used to be before intensive industrial agriculture drove dramatic soil carbon depletion. Technically, one might therefore question whether such activities really constitute NETs, given that the carbon had first been lost by emission into the climate system. Regardless, they are commonly ranked as NETs because the loss has occurred over long timescales and/or a long time ago, especially in the case of a lot of the deforestation. I will proceed under this caveat.

The second stream of NETs in this category centers around biochar. Carbon drawdown is achieved in this NET through widespread formation of biochar, notably using highly efficient new techniques that will be discussed. This biochar is highly resistant to decomposition and can be buried to achieve net carbon removal from the climate system over long periods of time.

The third stream of NETs in this group is known as enhanced weathering. It employs natural rock weathering reactions that draw $CO_2$ from the climate system and aims to accelerate these reactions in a bewildering variety of ways, to enhance the $CO_2$ drawdown capacity. An offshoot of enhanced weathering is the running of enhanced weathering processes in reactors under tightly controlled conditions, using the end-products to manufacture profitable goods.

## 4.2.1. Vegetation and Soils

In the most basic of terms, the growing of trees, plants, and algae draws down carbon dioxide levels because photosynthesis takes in $CO_2$ and separates it into carbon and oxygen. The carbon is used for growth and the oxygen is expelled into the atmosphere, giving us animals something useful to breathe in. Thus, growth of photosynthesizers results in carbon drawdown. But then most of the growth dies and decomposes on relatively short timescales, and this decomposition consumes oxygen and produces carbon dioxide. On a seasonal timescale, the so-called fluxes of carbon uptake and release are enormous—about 60 GtC per year each—but the close balance between carbon uptake during the growth season(s) and carbon release during decomposition leaves little to no net effect on carbon dioxide levels. If we want to use plant and algal growth as a NET, then we will have to find ways to increase growth, reduce decomposition, or both.

Trees are of specific interest in the NETs debate because many species live for decades, some for centuries, growing and capturing carbon all the while. Trees, and especially forests, thus have obvious and instant appeal when we are searching for longer-lived means of carbon removal. When considering trees at a whole-forest level, rates of carbon accumulation often appear to be optimal in young to middle-aged trees, tailing off at older ages. Detailed evaluation on the level of individual trees reveals a different picture, however, in which growth rates continue to increase with size and age in 97% of tropical and temperate tree species. Work in tropical wet forests of Suriname has supported this continued high or even accelerating carbon capture in very old trees. It is therefore a cause of grave concern that big old trees are in rapid decline on a global scale. To put it simply: protecting ancient forests likely is more efficient for carbon drawdown than replacing them by new forest. But this does not mean that all we should do is preserve old growth. Reforestation where tree-numbers are declining and afforestation or development of new forest in previously barren locations remain important additional solutions.

From a NETs perspective, it is critical to remember that even big old trees don't live forever. Inevitably death comes, and decomposition then returns the carbon back into the environment. A large portion re-enters the atmosphere as $CO_2$, and some as methane. To allow tree growth to be counted as an effective long-term NET, the fraction of greenhouse gases returned into the atmosphere has to be smaller than the fraction of greenhouse gases removed from the atmosphere during growth. And this net sum should be calculated in terms of total global warming potential because any carbon atom returned into the atmosphere in the form of a methane molecule has a much higher global warming potential than any carbon atom that was removed in the form of a $CO_2$ molecule.

Because a lot of the visible biomass of even big old trees normally gets recycled within decades of their demise, tree growing is often viewed as only a temporary means of carbon dioxide removal, dismissing the use of tree growing as an effective long-term NET. This dismissal hinges on a flawed perception of trees as mere lumps of wood above the ground surface. In reality, much more needs to be considered. Ecosystem productivity and carbon storage depend on a complex of interactions between processes such as photosynthesis, respiration, bedrock weathering, and soil formation. Trees have particularly long-lasting impacts on microbially produced organic matter that can survive in soils for millennia, and which facilitates the transfer of considerable amounts of a tree's carbon to neighboring trees and shrubs through root-microbial networks. The root-microbial networks also greatly accelerate chemical weathering of minerals, a process that drives further drawdown of atmospheric $CO_2$ on timescales of many millennia. In addition, a portion of dead tree matter itself gets buried in the soil, where specific organic compounds that are not easily decomposed can be preserved for many centuries to millennia. As a result, carefully executed reforestation and afforestation—with attention to re-establishing diverse ecosystems and soils—constitutes a potential NET with timescales that extend over many centuries or even longer. Moreover, such work around preserved ancient woodlands will both help the old growth to survive and thrive, drawing down more carbon dioxide into the big old trees, and help with providing critical soil ecosystem components to the newly established growth.

Growing new trees may not always mean active replanting. Many ecosystems that have suffered deforestation contain remnant seeds and root stocks in the soil, and natural regrowth from those resources can be cheaper and more successful than tree planting. In the Sahel, for example, a successful and fast landscape restoration technique has been trialed under the name of farmer-managed natural regeneration, through which subsurface remnant root stocks from trees that disappeared long ago are nurtured back to life. Major successes have been recorded in Tanzania (e.g., Barrow, 2014), and other work is ongoing on the Chinese Loess Plateau, on the Qinghai-Tibetan Plateau, and throughout arid Africa (Figure 4.1). This work has so far largely focused on landscape and ecosystem regeneration, and thus on improving habitability. But the potential benefits for $CO_2$ drawdown from the same efforts are also obvious and will be discussed later in this section.

The total NETs potential of reforestation, afforestation, and ancient woodland preservation depends on the combined net carbon drawdown capacity. A recent study determined that, beyond existing trees and agricultural and urban areas, there is room for a further 0.9 billion hectares of canopy cover under current climate conditions (Bastin et al., 2019). Smith (2016a,b) had estimated land requirements for afforestation and deforestation between a realistic 0.32 billion hectares and a grand maximum of 0.97 billion hectares, in agreement with the maximum potential estimated by Bastin et al. (2019). About 0.9 billion hectares would suffice for adding up to 1.5 trillion trees to the 3 trillion trees that exist on Earth already, which could store 205 Gigatons

**Figure 4.1**  Top: typical farm scenes before (a) and after (b) farmer-managed natural regeneration in Niger (World Vision Australia, 2019). Bottom: before (c) and after (d) landscape regeneration on the Chinese Loess Plateau (reproduced with permission from Liu and Hiller, 2016; see also The Environmental Education Media Project and John D. Liu, in Wahl, 2018).

*Source*: (a, b) Used with permission, Klezer.Gaspar@worldvision.com.au. https://fmnrhub.com.au/projects/niger/#.XuMf9C2r0Q9; (c, d) Used with permission, Rightlink https://medium.com/@designforsustainability/human-and-planetary-health-part-iv-restoring-ecosystems-in-the-century-of-regeneration-d24bbfe37617

of carbon under current climate conditions. But note that this storage value is sensitive to climate change; if we maintain the current climate trajectory, then the estimate reduces to only 150 GtC by 2050. If we take the middle of the two storage estimates and assume a timescale of about 100 years for mature forest systems to establish themselves, including their soil ecosystems, then we get a century-averaged carbon drawdown of roughly 1.75 GtC per year. This is similar to previous drawdown estimates of 1.8–1.9 GtC per year for combined reforestation and afforestation with soil carbon restoration (Smith, 2016a,b; Fuss et al., 2018). Although these numbers may seem large, they in fact are not even one fifth of the 10 GtC emitted by human activity each year. To complicate matters, continued climate change can alter these forest storage terms through heat and water-cycle changes (e.g., Lyra et al., 2016). The estimates discussed here are estimates based on current conditions.

To provide some context for the potential maximum of 205 Gigatons of carbon storage through reforestation and afforestation, let's consider the carbon in existing forest systems. The global carbon stock in today's forests is approximately 861 GtC, of which 44% is found in the top 1 meter of soils, 42% in living biomass above and below the surface, 8% in deadwood, and 5% in litter (Pan et al., 2011). In terms of geographic distribution, the same study found that tropical forests hold roughly 55% of the global 861 GtC, while boreal forests hold 32% and temperate forests 14%. Boreal forests are forests in high-latitude environments, with temperatures at or below freezing over 6–8 months of the year, and temperate forests occupy the temperate mid-latitude regions between the tropical and boreal regions. Use of the modern surface areas of these forest types to calculate carbon densities per hectare gives roughly equal numbers of about 240 tC per hectare for tropical and boreal forests. In tropic forests, 56% of carbon is stored in biomass and 32% in soil. In boreal forests, 20% is stored in biomass and 60% in soils. The carbon density in temperate forests is about 155 tC per hectare.

One thing is evident: consideration of forest-based solutions must include the soils that are integral and essential to these systems; almost half of the carbon storage potential of forests hinges on (re-)establishing healthy soil ecosystems. Given that monocultures tend to deplete rather than enrich soils and biodiversity, planting of monoculture forests for regular clearcutting does not seem to be the most promising way forward when trying to use tree growth as a NET. Instead, healthy soil ecosystems and biodiversity need to be stimulated through careful mixed-species regrowth and only partial harvesting. This might, for example, provide a near-continuous source of fuel for biochar production.

So, what's the status of tree cover on Earth? Is it on an upward trajectory or still on a trend of net deforestation? Satellite-based assessment suggests that tree cover has increased by 2.24 million square kilometers or about 7% between 1982 and 2015, despite ongoing deforestation, fires, and mortality caused by droughts and insect infestations (Song et al., 2018). The net gain was driven by expansion of tree cover in subtropical, temperate, boreal, and polar regions, which outweighed deforestation in the tropics. The analysis also revealed that the net gain in tree cover resulted from abandonment of agricultural lands in parts of Europe, Asia, and North America, from warming-driven poleward expansion of forests in response to warming, and from a massive tree planting program in China. However, tree cover does not necessarily imply proper forest cover; it also includes trees in industrial timber plantations, mature oil palm estates, and other planted cover. Hence, clearing ancient forests and replacing them with monoculture plantations would not show up as a net change in this tree cover assessment. But biodiversity, ecosystem services, and soil health and carbon retention would be affected detrimentally regardless.

This brings us to a key downside with respect to overenthusiastic afforestation or reforestation: not all tree cover is great for biodiversity and soil health, and therefore not all tree cover is great for carbon capture. An example is ongoing in Ireland, where a well-intended response to climate change has teamed up with a massive drive for commercial wood provision. The scheme aims to achieve net-zero carbon emissions by 2050 while offsetting rising emissions from agriculture and relies on planting 8,000 hectares of new forest each year, using fast-growing Sitka spruce that will be harvested for pulp, plywood, pallets, fencing, garden furniture, and building materials. Originally from North America, Sitka spruce thrives in the Irish climate and soil conditions and grows to maturity within 30 years, by which time it can reach over 100 meters tall. Its growth is exponentially faster than that of native oak. Unfortunately, most of the Sitka spruce planting comes in the form of tightly packed monocultures. These risk suffocation of biodiversity at the forest floor by densely packed spruce needles, as well as soil degradation and associated problems with carbon retention.

Albedo, or reflectivity change is another source of concern (Fuss et al., 2018). Afforestation of boreal forests, especially, can work counterproductively, because boreal forest has low albedo with respect to incoming sunlight. This effect will dominate over the impact of carbon drawdown regionally, and the end-result will be regional warming and accelerated loss of snow and ice. In temperate regions, the overall impact is close to zero, but will again be counterproductive if afforestation concerns temperate forest in place of agricultural land, which has a relatively higher albedo. It is mainly in the case of tropical afforestation that the effect of carbon drawdown will dominate over albedo changes.

A further issue with a focused drive toward reforestation and afforestation is that ancient grassy biological systems, or biomes, may too easily be identified as opportunity areas for tree plantations and forest expansion. This would overlook the critical environmental roles played by such ancient grasslands, savannas, and open-canopy woodlands in maintaining biodiversity and so-called ecosystem services. Ecosystem services are an ecosystem's direct and indirect contributions to human well-being, including climate regulation, water purification, pest and disease control, and cultural aspects. Transforming healthy grassy biomes into forests, and especially into monoculture plantations, comes at high cost to biodiversity and ecosystem services. In addition, healthy, non-degraded grassy biomes have a carbon storage potential up to two-thirds that of healthy, varied forests, albeit with as much as 85% of the carbon inside the soil and only 15% above the surface. Expanding or planting monoculture forests in replacement of ancient grassland will therefore do more damage than good. Biodiversity and ecosystem services will be degraded, and the area's carbon storage potential will likely be reduced as well. Originally stored carbon may even be released into the atmosphere as $CO_2$.

Conversion of dry grassland areas to agriculture, or any other degradation of the grassy biomes, similarly reduces the carbon storage potential, especially in arid zones of the world. And grasslands that become overused in terms of plant production may again turn into net sources of $CO_2$ release into the atmosphere. Clearly, careful protection of both ancient forests and grasslands is of critical importance, given that both comprise well-established processes of carbon drawdown and contain much stored carbon that would be released as $CO_2$ (or methane) if these ecosystems are degraded.

\* \* \*

The historical development of intensive, industrialized agriculture has caused widespread ecosystem and soil degradation (Figure 4.2), resulting in a major net flow of soil carbon into the climate system. This is partly why agriculture counts as a substantial net $CO_2$ emitter. In places, soil degradation has been so intense that soil erosion now takes place on massive scales; examples include the Chinese Loess Plateau, large tracts of Australia, major portions of the central United States, among many other places. Loss of organic matter from soils has severely reduced their capacity to retain water and loss of living, intertwined root systems has strongly diminished soil cohesion. As a result, soils get washed away by every rainstorm, when hardly any of the falling water can be retained in the soil, or blown out in large dust

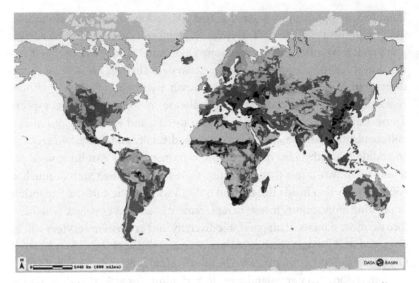

**Figure 4.2**  Soil degradation map of the world (increasingly dark tones indicate increasingly degraded soils).

*Source*: Created by Wendy Peterman; used under Creative Commons Attribution 3.0 License (data from Oldeman et al., 1991). https://databasin.org/maps/58940a97a23341db 8233844a2be14319/active. For other soil degradation maps, see Middleton et al. (1997).

storms during every dry period. These trends can be reversed by so-called regenerative farm practices, which are closely related to the farmer-managed natural regeneration we have encountered before in this section.

Regenerative farming aims to create sustainable productivity through efficient use of resources and stimulating natural cycles, processes, and ecosystem interactions. It often includes measures such as landscape terracing or contouring to help water retention; improved irrigation and water cycling, including measures to limit evaporation by trickling water through pipes rather than flowing it through open ditches, and covering water reserves; mixed cropping and crop rotation to stimulate natural nutrient cycles; planting of trees to provide and maintain shade, root networks, and microbial ecosystems; intensive pasturing to encourage natural breaking up of the soil surface by hooves and fertilization by excrement; limiting the use of industrial fertilizers and insecticides, and stimulating resilience through development of natural, balanced and diverse ecosystems; and so on. On the Chinese

Loess Plateau, severely degraded environments that were turning into desert were spectacularly restored in this way and now produce good quantities of food. In Australia, regenerative farming pioneers are realizing considerable productivity increases on land that had previously become marginal, and similar results have been reported from Africa and the Americas.

Landscape regeneration can in principle be commenced at scale and immediately, but while humanity globally can be kept busy for a long time with the regeneration of previously degraded lands, carbon drawdown into vegetation and soils is not an indefinite solution from a systems point of view. The vegetation and soil reservoirs will eventually reach a saturation point, when they are "full," and their annual $CO_2$ removal will then reduce to zero. Thereafter, a steady cycling system may be maintained, with high carbon stocks, but with a carbon flux into the system that will closely balance the carbon flux out of the system. From a NETs perspective, vegetation and soils will thus be effective during the active enrichment phase, but will become negligible once saturation is reached. There is a risk of carbon loss from restored ecosystems because of calamities, such as floods, droughts, fires, pests and diseases, or poor future management (IPCC, 2019), and general warming of forests and soils may also increase decomposition rates and carbon loss. Yet, all of these problems would also affect target regions if no regeneration were undertaken and might then have even greater impacts. This is because regenerated landscapes will have more built-in resilience to such problems than non-regenerated landscapes, owing to their improved qualities of water retention and ecosystem diversity.

Naturally, there are also concerns about widespread landscape regeneration. One concern is that excessive water capture could cause downstream water scarcity, which would need to be carefully monitored and addressed. Water requirements are very high for afforestation and deforestation. They have been estimated at a realistic 370 km$^3$ per year, although some estimates reach a maximum of 1040 km$^3$ per year (Smith, 2016a,b). However, water quality is important too, and landscape regeneration is known for its capacity to improve downstream water quality by reducing soil erosion into waterways and thus downstream silt-loading of the water. For example, the sediment flow from the Chinese Loess Plateau into the Yellow River has been reduced by more than 100 million tons per year, which has substantially improved the water quality.

Worries have also been expressed about the longevity of (parts of) the regeneration projects, as exemplified by high mortality of newly planted Chinese pine trees on the Loess Plateau, which may not have been the most suitable species for the prevailing semi-arid conditions. Cultural and community involvement in regeneration initiatives is important too, for ensuring that the community is fully behind the landscape modification and species selection for planting. This community involvement is also vital for long-term acceptance and maintenance. Thus far, regenerative agriculture has largely focused on re-establishing farm productivity in marginal settings and on reducing the use of industrial chemical fertilizers and insecticides. The substantial apparent increases in carbon fixation in plants and especially soils have so far played second fiddle. If scientific monitoring confirms that carbon is indeed stored and locked away in soils in useful quantities over multiple centuries, however, then regenerative farming becomes a potentially critical NET that uniquely combines carbon drawdown with food production. This places it in stark contrast with intensive, industrial agriculture, which emits carbon during the process of food production. Eventually, a national and/or international carbon price will be agreed, either as a punitive cost of carbon emission, or as a reward or credit for carbon removal (or both). When used as a credit, it would open an exciting new revenue stream for regenerative farmers, supporting the returns they get from their farm products.

And then there is the biggest discussion-point of all. This is the serious risk of all-out competition for everything including land area and resources (notably water and nutrients) between four main players: reforestation and afforestation, native ecosystem conservation, farming for bioenergy crops, and farming for food production. Food production for humans needs to approximately double by 2050. More land and resources will likely be needed to achieve this, although some relief may come from technological solutions such as development of high-intensity crops and microbial protein. But conservation of ancient woodlands and grasslands, and expansion of such ecosystems (or similar) for carbon drawdown, also require land area and resources. Biomass-based renewable energy and fuels, biochar and PyCCS (pyrogenic carbon capture and storage), and especially BECCS (Bio-Energy with Carbon Capture and Storage) applications for energy generation with carbon removal (sections 4.4.2 and 4.4.1) will place even more, potentially massive demands on land and resources.

In more specific terms, a comprehensive recent assessment concluded:

> We find that the large-scale use of bioenergy, if not implemented properly, would raise food prices and increase the number of people at risk of hunger in many areas of the world ... This risk does not rule out the intensive use of bioenergy but shows the importance of its careful implementation, potentially including regulations that protect cropland for food production or for the use of bioenergy feedstock on land that is not competitive with food production. (Hasegawa et al., 2020)

On a more positive note, targeted, region-specific approaches to the planting of trees and grasses can also benefit agriculture through improved soil and water management (Hambrett, 2018), which suggests that the solution may in fact lie in achieving synergies between the two competing interests.

To navigate the conflicting interests for space, water, nutrients, and even labor and equipment availability, a sound trans-national framework of agreements will be needed. No matter how well intended, unilateral actions without such oversight could cause seriously detrimental impacts on other regions or nations, or on another sector within the same region. Consider, for example, how upstream construction of a major dam and irrigation infrastructure might cut off essential water supplies for food and drinking-water production in downstream regions or nations. It is clear that humanity cannot risk interruptions to food and water provision for the sake of carbon drawdown. But equally, humanity cannot risk uncontrollable climate change because of insufficient carbon drawdown, especially given that this climate change would in turn jeopardize food production. A balanced, internationally agreed approach is needed, with continuous monitoring and, if necessary, adjustment.

In arid, fire-prone regions like Australia, landscape regeneration has additional advantages in that it helps to restore better soil humidity and thus water availability in creeks and ponds as well as a reduction in the abundance of tinder-dry fuel for bushfires. Moreover, enhanced soil moisture and vegetation help to establish a more active hydrological cycle, increasing air humidity and potential for cloud and rain formation. Any reduction in bushfires is valuable in limiting this cause of massive human and animal suffering, environmental devastation, and release of $CO_2$, harmful air pollution, and toxins.

## 4.2.2. Biochar

Next in line for discussion on land-based NETs is biochar. Biochar can be made from many types of biomass, such as trees, including residues after tree harvesting for timber, crop waste, grasses, and manure. Biochar can be used in a variety of ways, including fuel development, burial for long-term carbon drawdown and soil enrichment, and suppression of soil emissions of methane and nitrous oxide (see also Chapter 3). Biochar addition to soils can help with water retention and nutrient enrichment and drive substantial crop-yield improvement, especially in regions with acidic soils.

The Terra Preta in the Amazon rainforest demonstrates that effective carbon burial may be achieved with biochar over many centuries to millennia. Moreover, the carbon in biochar does not generate methane if landfilled or added to soil, and biochar added to soil or compost can reduce methane, $N_2O$, and $CO_2$ production from soil. Overall, biochar's global carbon drawdown potential was estimated by Smith (2016a) at about 0.7 GtC per year, although other studies suggest only about half that value. Smith (2016a) in addition says that biochar requires between a realistic estimate of 0.04 billion hectares to a very maximum of 0.26 billion hectares in land area, while its water requirements were found to be negligible. Extensive biochar experiments are ongoing across the world already (Scholz et al., 2014), and there is nothing to stop the immediate implementation of large-scale biochar production and admixture to soils. However, not all relevant questions have been answered yet; control experiments and monitoring remain necessary to ensure optimization of results along with minimization of adverse effects.

Four-year tests with straw-derived biochar addition to rice-culturing soils in the central subtropical region of China revealed that methane emissions from the field were significantly reduced and total soil carbon content had increased, although $N_2O$ emissions were enhanced. Overall, there was a net reduction in terms of global warming potential because the influence of the methane reduction and soil carbon increase dominated over that of the increase in $N_2O$ emissions (Wang et al., 2018). The study also found that the biochar addition improved soil fertility by decreasing soil acidity and increasing total soil organic carbon, nitrogen, and phosphorous. Moreover, other studies have found that $N_2O$ emissions dropped when biochar was applied. In any case, the $N_2O$ effect seems short-lived, whichever way it operates (Woolf et al., 2018).

The pyrolysis process by which biochar is formed from harvested biomass produces energy, which can be used in replacement of fossil-fuel energy. Pyrolysis also produces gases that are collectively known as pyrogas. When cooled, part of the pyrogas condenses to form liquid bio-oil, while part remains gaseous and is known as permanent pyrogas. In most current biochar production methods, bio-oil and permanent pyrogas are combusted, releasing $CO_2$ to the atmosphere. If the pyrogas is instead put to good use as a source of energy or biofuel, then its combustion at least partly avoids the combustion of fossil fuels. Given that biochar-derived energy or fuel would be emissions neutral, its use technically falls under emissions avoidance rather than NETs.

A new method, called pyrogenic carbon capture and storage or PyCCS, aims to use the entire process as a complete NET system, rather than only the biochar component (Figure 4.3). For this, it includes carbon removal for all three pyrolysis products through use as bio-based materials and agricultural supplements, and through carbon storage in geological reservoirs (Schmidt et al., 2018). It is estimated that PyCCS can deliver efficiencies of drawdown and long-term storage of carbon that exceed 70%, versus 30–50% for regular biochar formation. Hence, the total potential impact of PyCCS would be to roughly double the biochar-only value of 0.7 GtC per year to about 1.4 GtC per year, while the land surface-area requirement would remain the same.

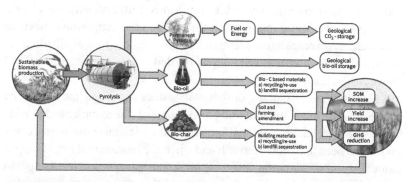

**Figure 4.3** Overview of the PyCCS process.

*Source*: Following Schmidt, et al. Pyrogenic carbon capture and storage. *GCB Bioenergy, 11,* 573–591, 2019. https://doi.org/10.1111/gcbb.12553

There are distinct advantages to biochar production and especially PyCCS. After processing, a substantial portion of the permanent pyrogases consists of hydrogen and $CO_2$, which could be used as a zero-emissions fuel and stored long term in geological reservoirs, respectively. Further advantages are that the technology of pyrolysis is well understood already, biochar and bio-oil storage in soils and biomaterials can be done without ecological hazards, and global scale-up of the PyCCS approach seems feasible within 10–30 years. Biochar and PyCCS therefore offer critical early opportunities for carbon drawdown at large scales, buying us time for spinning up other, currently less well-developed NETs.

Biochar addition to soils affects not only carbon, nutrient, and water cycling, but also the reflectivity or albedo of the soil. Specifically, it darkens the soil—which, incidentally, compost does also—so that slightly less sunlight (ISWR) is reflected back into space. This has a warming effect that would be opposite to what we would be aiming for with biochar addition. In consequence, it is critical that we understand whether the albedo effect could overwhelm the avoided $CO_2$ emissions effect of biochar—if it could, then we'd need to think twice about adding biochar to soils. Woolf et al. (2018) evaluated these impacts and concluded that the theoretical influence of 5% biochar enrichment would impart a positive radiative forcing equivalent to about 1% of the negative radiative forcing from avoided $CO_2$ emissions that is achieved with the same biochar application. Including observations from other studies as well, they find that this offset could be as high as 30% under unfavorable conditions and as low as zero under favorable conditions. But, crucially, the albedo effect was never found to outweigh the avoided $CO_2$ emissions effect. In sum, there is no reason to reject biochar addition because of albedo concerns, although careful monitoring is needed to ensure that the optimal returns are achieved.

Finally, full understanding of these NETs does not hinge just on the net full-lifecycle carbon emissions and albedo changes due to dark biochar addition to the land. Rather, potential landscape albedo changes due to tree cover would also need to be considered, if and when the biochar and PyCCS applications rely on afforestation. This may impart counterproductive changes in the regional radiative budget of climate, and thus temperature, as was discussed in the section dealing with afforestation (4.2.1). Note that this applies also to the NET known as BECCS (section 4.4.1).

## 4.2.3. Enhanced Weathering

The third major family of land-based Earth System NETs concerns enhanced chemical weathering of rocks during which $CO_2$ is consumed. This stream of approaches focuses especially on weathering silicate minerals like olivine (whose composition ranges between $Mg_2SiO_4$ and $Fe_2SiO_4$, with Mg dominating). Olivine forms part of, and in some cases almost completely dominates, high magnesium and iron bearing rocks such as basalt, gabbro, dunite, diabase, and peridotite. In particular, dunite is a type of peridotite that comprises more than 90% olivine. These rocks and minerals are highly susceptible to weathering in the Earth's surface environment.

The weathering of olivine is described as $(Mg,Fe)_2SiO_4 + 4CO_2 + 2H_2O \rightarrow 2(Mg^{2+},Fe^{2+}) + 4HCO_3^- + SiO_2$. The reaction requires water and $CO_2$. It produces magnesium and iron ions as well as bicarbonate ions, all of which are transported away dissolved in water. $SiO_2$ is commonly known as silica, and a considerable part of this component will also be dissolved in water as silicic acid ($H_4SiO_4$). In practice, magnesium-carbonate minerals and iron oxides may form from the reaction products. The formation of magnesium carbonates will re-release up to half the $CO_2$ that was taken up during olivine weathering and fix the other half in the carbonate minerals for millions of years. Where the dissolved reaction products reach waterways and eventually the ocean, they add so-called alkalinity—the opposite of acidity—which results in greater $CO_2$ dissolution from the atmosphere into the water. Reaction products may also be distributed into the ocean via pipelines or ships, or the reaction may take place directly in the ocean, for example through enhanced weathering on beaches (Figure 4.4).

For summaries of olivine and its potential by a key pioneer in the field—and, incidentally, one of my university lecturers—see Schuiling and Krijgsman (2006) and Schuiling (2019). Field observations of geological outcrops indicate that olivine and olivine-bearing rocks such as dunite weather very rapidly. In these natural environments, with intensive microbial action, dunite and olivine weathering rates are about 10 times higher than in sterile laboratories. Field trials with olivine added to soils have revealed intermediate weathering rates. Higher temperatures speed things up even more, so that weathering rates in the tropics exceed those in high latitudes. Under favorable conditions in the humid tropics, fast penetration of weathering may be expected at up to 50 micrometers per year.

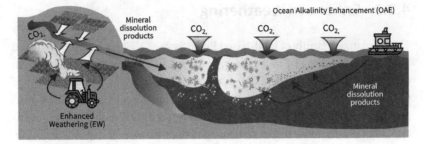

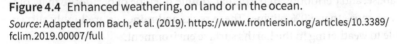

**Figure 4.4**  Enhanced weathering, on land or in the ocean.
*Source*: Adapted from Bach, et al. (2019). https://www.frontiersin.org/articles/10.3389/fclim.2019.00007/full

For scale, removal of all $CO_2$ from the atmosphere would require complete reaction of an amount of olivine equivalent to a 0.4 cm layer spread over all of the world's land mass. But this ignores the fact that once atmospheric $CO_2$ starts to be drawn down, the ocean would start to outgas $CO_2$ that had been taken up when atmospheric levels were rising. If we include that effect, then the required olivine layer is closer to 0.7 cm thick. There is no shortage of olivine rocks even for that amount, and we don't want to remove all $CO_2$ from the atmosphere anyway. Drawing down some 20% of it would bring $CO_2$ concentrations into a sustainable range below 350 ppm, and this reduces the required thickness of our hypothetical olivine layer to 0.15 cm, assuming that it all weathers perfectly.

Weathering such amounts of olivine would take many decades to—more realistically—centuries, even under favorable conditions. To be considered an effective NET, timescales of decades would be acceptable, but timescales of centuries make a process much less useful. So, the olivine rock weathering process needs to be accelerated. This brings us to the concept of enhanced weathering: the addition of olivine rock to environments, either natural or manipulated, in which the material is subjected to increased rates of weathering.

One option is to place fine olivine grains in environments where they grind into each other, such as running water or waves. This continuously removes the weathered silica coatings, exposing fresh olivine for further weathering and can be done both in freshwater and seawater environments. Coastal application is promising because olivine reacts particularly well with sea water, because there is a lot of wave energy, and because the released products are all natural and harmless, especially in diluted form in the ocean. The released silica is a crucial growth nutrient for silica algae such as diatoms, which grow fast and are rich in fatty acids. Diatoms are good fish food and also have strong potential for biodiesel production.

Applied to soils on land, ground olivine rock has potential for improving soil structure and water management. Its weathering not only binds $CO_2$ but also neutralizes soil acidity and provides nutrients. As we have seen before, biochar addition to soils brings similar benefits. There is now an increasing level of interest in the international community in combining olivine rock application to soils with biochar addition, to see how the two approaches for fertilization and $CO_2$ removal can complement and possibly intensify one another. Because so-called ultramafic olivine-rich rocks—the peridotites, including dunite, and also the hydrated olivine form called serpentinite—can contain high concentrations of chromium and/or nickel, accelerated weathering of these materials results in fast releases of biologically harmful nickel and magnesium-driven suppression of plant uptake of nutrient calcium. For on-land applications, therefore, basalt is commonly preferred over such rocks, given that basalt also weathers rapidly and releases many useful nutrients, such as potassium, phosphorus, calcium, magnesium, and iron. Depending on soil and crop types, an addition of 10–30 tons of fine-ground basalt per hectare per year to two-thirds of the globally most productive cropland soils (that is, 0.9 billion hectares) may achieve carbon removal rates from the climate system between about 0.1 and 0.9 GtC per year by 2100, after accounting for the $CO_2$ emissions associated with mining, grinding, transporting, and spreading of the basalt (Moosdorf et al., 2014; Beerling et al., 2018).

Incidentally, industrial waste such as cement and especially concrete from demolition, and slag from iron or steel manufacturing, are promising for enhanced weathering as well, in addition to basalt or olivine-rich rocks. However, while promising and definitely useful to pursue, these products do not really represent NETs, given that their production involved $CO_2$ emissions. Their $CO_2$ uptake at best represents emissions reduction back toward zero, rather than reaching into net negative territory, which would be needed to count as a NET. Similar restrictions apply to potential carbonation reactions of other waste products, such as coal fly ash, saline waste water, and municipal solid waste. Moreover, the amount of human-caused $CO_2$

emissions that can be removed using industrial products is relatively small, given that available amounts of such industrial products are much smaller than the amount of emitted $CO_2$.

A recent overview of enhanced weathering under a wide variety of conditions covers a range of source materials and techniques, and provides a wealth of references for further details (Romanov et al., 2015). Although this review addresses many different mineralogies, rock types, and industrial waste types, we will here mainly limit the discussion to olivine-rich serpentinite or basaltic rocks that are pivotal to discussions of enhanced weathering as a NET. Some of the key findings are that source-material particle sizes well below 0.1 millimeters are crucial for achieving effective carbonation. Reactions were also found to be sensitive to agitation of the particles, for removing reaction products and exposing fresh surfaces to weathering. Mild agitation was found to speed up leaching of cations (e.g., $Mg^{2+}$ or $Ca^{2+}$) from the rocks, although too much agitation hindered it by reducing the contact areas between grains. Moreover, higher rates of cation leaching were found at lower pH, which indicates increased acidity, and—conversely—increased deposition of new carbonate minerals at higher pH. Thus, the leaching of cations from source rocks and new carbonate deposition are best separated and maintained at different pH levels. This separation can be either spatial or temporal. Spatial separation implies that two different physical sites are set up for the two different processes, with transport of reaction products between them. Temporal separation is achieved using two successive treatment steps, adding first an acid and then a base; this has become known as the pH-swing method. The acid treatment phase can be associated with heating, with promising results found at temperatures up to 650°C.

Not all olivine-bearing, serpentinite, or basaltic rock applications on land require new mining and distribution of the material. There are many mining-waste heaps, known as mine tailings, which mostly contain such rocks. Developing ways to accelerate weathering of these mine tailings offers an attractive means of turning industrial waste materials into key resources for $CO_2$ removal from the climate system. For the mining companies, this can offer a way of monetizing something that was previously considered only a cost item. Acceleration of mine-tailing weathering may involve pumping hot water or steam with concentrated $CO_2$ through thick layers of tailings to ensure deep penetration, whereas normal weathering would reach only the uppermost layers. The weathering then produces alkaline conditions in the percolating water and permanent fixation of $CO_2$ in the form of newly deposited carbonate minerals.

While working with mine tailings might sound straightforward in general terms, the devil—as always—is in the detail. Mine tailings of different mineral compositions and different permeability give greatly different weathering rates. Fine-grained material is more reactive but also offers less gas permeability. Coarser-grained materials are less reactive but offer better gas permeability. Hence, achieving the best results will likely require development of case-specific co-disposal strategies of the fines and coarser materials (Zarandi et al., 2017). In terms of impact, careful assessment has demonstrated that reactions in tailings at the Woodreef mine site in New South Wales, Australia, have removed between 3900 and 6900 tons of $CO_2$ from the climate system over the period 1983–2013 (Turvey et al, 2018). That is 130–230 tons of $CO_2$ per year, or 35–63 tons of carbon per year. This pales to insignificance when compared to the human-caused emissions, which amount to about 10 billion tons of carbon per year. Based on the naturally occurring weathering rates determined at Woodreef, one would need similar reactions at 44–77 *million* similar mine sites only to offset human-caused emissions, or more if we want to enter into negative emissions territory. This underpins the growing interest in potential methods for enhancing the weathering rates.

The mafic and ultramafic rocks of interest are widespread around the world, and there is intensive mining in them, but not yet to the required scale calculated here. Hence, we can conclude that acceleration of mine-tailings reactions is a promising avenue for helping with carbon reduction, but that by itself, it remains far from a complete solution to the problem. There are also concerns about transport of potentially harmful compounds from mine tailings in the percolating fluids and eventually the groundwater. It is therefore essential that each potential case is carefully assessed for feasibility, environmental impacts, and potential quantity of $CO_2$ removal versus $CO_2$ releases associated with the processes needed to accelerate the weathering. To count as true NETs, a purist might say that the net $CO_2$ removal should also account for the $CO_2$ produced during generation of the mine tailings. But the tailings are generated in the extraction of otherwise needed materials, so that they and their associated emissions would be generated anyway. Hence, acceleration of mine tailing does count as a NET, because it utilizes a resource that would otherwise just be left and ignored as waste.

An alternative approach to accelerated weathering in reactors or waste heaps revolves around the pumping of concentrated $CO_2$ into wells drilled into olivine-bearing or basaltic rocks. Experimental results, again, differ somewhat from pure theory. Notably, different basalt formations—in ways that seem unrelated to their chemical compositions and mineralogy—display different reactivities to water-dissolved $CO_2$ or combinations of water-dissolved $CO_2$ and $H_2S$, where $H_2S$ is added to increase the acidity of the injected fluid (Schaef et al., 2010).

The first quantitative results from $CO_2$ injection into a basalt formation were obtained in two trials in Iceland, using 73 and 55 tons of $CO_2$ from the Hellisheidi geothermal power plant. This power plant emits $CO_2$ of volcanic origin as a byproduct, and there is also a collaboration with a facility that captures $CO_2$ directly from air (section 4.4.2). The $CO_2$ was dissolved in water within the well during its injection between 400 and 800 meters deep into the basalt. It was found that more than 95% of the $CO_2$ was immobilized between the injection and monitoring wells over a duration as short as 2 years, and this immobilization was attributed to carbonate precipitation (Matter et al., 2016).

A later study, however, provided a very different explanation of the $CO_2$ trapping, based on detailed analysis of changes in deep microbial ecosystems in response to the $CO_2$ injections. It found that bacterial biomass had increased 500 fold in the monitoring well, within two months of $CO_2$ injection (Trias et al., 2017). The biomass consisted of so-called chemolithoautotrophic bacteria, which—in plain speak—means that these bacteria likely drove $CO_2$ conversion into biomass, and thus played a major role in the rapid trapping of $CO_2$ (Daval, 2018). This biological trapping in biomass may not be of a similarly permanent nature as geological trapping in new minerals. Thus, major uncertainties arise with respect to the efficiency and longevity of $CO_2$ trapping by injection into basalts, which need to be addressed urgently in further studies.

Overall, the total potential of land-based enhanced weathering by 2050 is estimated to be about 0.5–1.0 GtC per year (Fuss et al., 2018).

Finally, there are also other methods for geologic carbon storage, in which pressurized $CO_2$ is injected into rock formations buried up to hundreds of meters below Earth's surface, including injection into active oil fields or into deep salty brines. Injection of $CO_2$ into active oil fields is used to enhance oil recovery and thus oil-based production of more $CO_2$. In consequence, it might at best be carbon neutral, and at worst will be conducive to adding new emissions. Therefore, this practice has no place in a discussion of NETs. Injection of $CO_2$ into deep salty brines relies on solubility trapping in the fluids. This may not be a long-term or permanent solution, although there may be a component of slow fixation by new mineral deposition. Longevity of storage remains a concern with this method and requires careful case-by-case assessment.

## 4.2.4. Products Made from Captured $CO_2$

Manufacture of products by mineralizing carbon dioxide relies on reactions of concentrated $CO_2$ with waste silicate minerals from mines (as in section 4.2.1) to produce products that have their own value and thus potential profitability (Figure 4.5). Thus, it has much in common with enhanced weathering and, in fact, has been inspired by those processes. In the $CO_2$ mineralization or mineral carbonation approach, however, the processes take place in reactors under carefully controlled conditions and with targeted separation of reaction products. The concentrated $CO_2$ may derive from captured and concentrated industrial emissions (section 3.4), including those from bio-energy plants (section 4.4.1), or from direct air capture (section 4.4.2).

$CO_2$ mineralization can use olivine-bearing, serpentinite, or basaltic rock as feedstock to the processes. Alternatively, alkaline industrial waste materials such as steel slag and fly ash might be used. Depending on feedstocks, the main output products are magnesium carbonate, metal oxides, and silica. These products are valuable for the manufacture of advanced cements, plaster boards, building aggregates, and concrete reinforcements; filler and reinforcing of plastics and rubber; and additives for pharmaceutical and food industries. There is specific interest in building materials because these are used globally in gigantic quantities, and they store carbon over timescales of decades to centuries. It has been estimated that replacing 10% of global building materials with $CO_2$ mineralization products could offer a realistic drawdown of 1.6 Gt $CO_2$ (or 0.44 GtC) per year (Gadikota et al., 2015), though other estimates for total carbon removal by using it to make products are lower, at around 0.2 GtC (Mac Dowell et al., 2017).

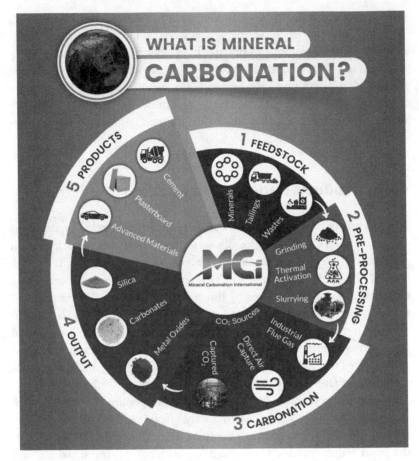

**Figure 4.5** Example of mineral carbonation.

*Source*: Used by kind permission, Mineral Carbonation International (MCi, 2020). https://www.mineralcarbonation.com

$CO_2$ mineralization can operate with much lower concentrations of $CO_2$ (roughly 20%) than the high $CO_2$ levels (preferably 100%) typically sought for storage in deep geological formations. These lower required $CO_2$ concentrations are comparable with the gas concentrations seen in outlets from cement kilns and only slightly higher than those from coal-fired plants (Litynski, 2019). Hence, the requirements of $CO_2$ purification are relatively low for $CO_2$ mineralization, which reduces energy demand.

There are substantial downsides to consider too. Most importantly, $CO_2$ mineralization in isolation is not a complete NET, as it relies on a supply of concentrated $CO_2$ from other processes. To make it into a NET, therefore, a combined application needs to be set up that includes $CO_2$ capture and subsequent mineralization. Even if the $CO_2$ supply is taken for granted, recent assessments indicate that $CO_2$ storage via $CO_2$ mineralization in reactors remains significantly more expensive than $CO_2$ storage in deep sedimentary formations (National Academies of Sciences, Engineering, and Medicine, 2019). The same report highlights that global upscaling of the process raises issues around transporting and storing Gigatons of carbonated solid material, with potentially large costs and yet unknown environmental impacts, and greenhouse gas emissions. Yet, these objections seem to assume that the mineralized products are just for storage, where in many cases costs may be offset by the commercial value of the products that are produced.

In any case, however, questions remain as to the amount of $CO_2$ reduction that could be achieved by making products of commercial value. Globally, humanity uses about 100 billion tons of building materials, about half of which is sand, clay, gravel, and cement for building, plus minerals quarried for fertilizer (Carrington, 2020). If we make a bold assumption that up to half of that (25 Gt/y) could be replaced in the most favorable case by mineralized carbon, then a mix of magnesium and calcium carbonates of 25 Gt/y would represent roughly 3.3 GtC/y. Thus, in our most favorable (most would say unrealistic) scenario in which mineralized carbon production is scaled up immensely to a point where it replaces half of the world's sand, clay, gravel, and cement use, and in which this production process itself would bring no new emissions, then carbon mineralization has a NET potential of at most one-third of current carbon emissions. However, realistic estimates converge on about 0.40 GtC/y, at least for the foreseeable future (see section 4.6); though the process remains important as part of a wider portfolio for $CO_2$ removal, especially because it represents a potential growth sector that could sustain itself commercially.

## 4.3. Marine Earth System NETs

Similar to Earth System manipulations on land to drive $CO_2$ drawdown, such manipulations are possible in seas and oceans as well. The difference is that marine applications are even less mature in their development. Theoretical arguments exist, and in some cases quite good modeling assessments, but true field-trial experiments are few and far between. Two specific developments pose international legal limitations to carbon drawdown research (Brent et al., 2018).

First, in 2010 the UN Convention on Biological Diversity (CBD) invited all countries and international organizations to prohibit geoengineering activities (including research) that may negatively impact biodiversity until such time that science can justify geoengineering trials through a better understanding of risks and impacts. Small-scale scientific research in controlled settings is permitted, but that excludes trials in the open ocean or the atmosphere. The prohibition was re-affirmed in 2012 and then in 2016. Although non-binding, this prohibition is important because it sets the community's tone relative to geoengineering research.

Second, in 2013 the London Convention and Protocol—which internationally governs the dumping of waste at sea—adopted Resolution LP.4(8). This resolution concerns a legally binding prohibition on ocean fertilization (although it is yet to come fully into effect). It was imposed in response to controversial proposals for ocean fertilization that were aimed at creating emission reduction credits for sale on emerging carbon markets. Again, there is an exception for scientific research, but only if it adheres to a detailed set of criteria. And here's the catch: ocean fertilization research activities are prohibited if their design, conduct, and/or outcome is influenced by economic interests, or leads to direct economic gains. Brent et al. (2018) find that: "at face value, the London Protocol prohibits any ocean fertilization research that is aimed at or leads to economic gains, such as carbon trading, an approach that may be applied to other marine geoengineering research in the future." In other words, this stricture may not just apply to ocean fertilization; it might equally be used to limit other marine NETs research that includes addition of substances to the ocean.

In spite of the issues with respect to large-scale field testing and implementation, four major streams of marine Earth System NETs merit specific mention (Figure 4.6). The first is ocean fertilization, in which ocean patches are seeded with nutrients that are locally in short supply to trigger enhanced plankton blooms that then—in theory—result in the sinking of more organic matter into the deep sea and onto the sea floor where it might get buried. The second, the so-called blue carbon approach is somewhat related and focuses

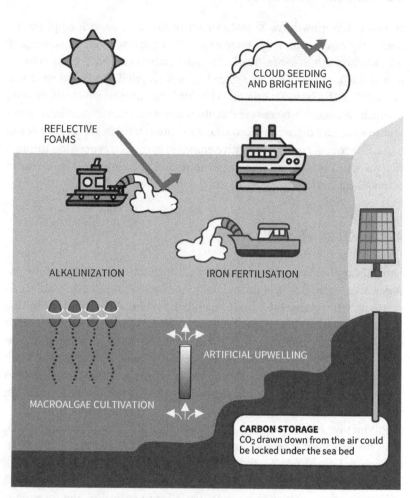

**Figure 4.6** Marine-based approaches for so-called geo-engineering of the climate. Cloud seeding and ocean foam are not NETs, but reflectivity enhancers, which will be discussed in Chapter 5.

*Source*: Simplified following GESAMP. High level review of a wide range of proposed marine geoengineering techniques, *GESAMP Reports and Studies*, *98*, 143, 2019. https://www.nature.com/articles/d41586-019-01790-7

on vegetation growth (re-)establishment in the oceans, with emphasis on coastal regions. It especially concentrates on mangroves, kelp (macroalgae), seagrass, and salt marshes. The key to blue carbon as a NET is the component of carbon that is actually stored, versus recycled, similar to what was discussed in the previous section. The third stream is ocean alkalinization, in which alkaline substances are distributed into the oceans to counter ocean acidification and drive enhanced $CO_2$ absorption into the alkalinized ocean water. The fourth concerns negative emission hydrogen generation through electrolysis of sea water and use of the residual alkaline brine to perform ocean alkalinization.

## 4.3.1. Ocean Fertilization

Limited supply of any of the major or trace nutrients required for photosynthesis acts to restrict algal production because photosynthesis requires all nutrients to be present in a certain balance. By artificially adding the limiting nutrients, ocean fertilization aims to trigger algal blooms (Figure 4.6), which develop in surface layers where sufficient light penetrates for photosynthesis; that is, the upper 100 meters or so. Evidence that such fertilization can work comes from the observed impacts of falls of nutrient-rich volcanic ash onto the ocean and from limited field-trials in which iron was released, mainly in the form of iron sulfate.

Production of algal organic matter through photosynthesis draws $CO_2$ from the dissolved inorganic carbon reservoir in surface waters, notably in the forms of dissolved $CO_2$ and of bicarbonate, or $HCO_3^-$ (Rost et al., 2003). The underlying reaction is given by $CO_2 + H_2O \rightarrow CH_2O + O_2$, where $CH_2O$ indicates organic matter in its most basic form; in reality, the formula is much more detailed, but we don't need that level of detail here. The surface-water inorganic carbon reservoir is in continuous gas exchange through the sea surface with the atmosphere. The idea behind using ocean fertilization as a NET is that, upon death, the triggered algal blooms sink to deep waters and eventually the sea floor, where burial of the organic carbon would constitute a net removal of carbon (and thus carbon dioxide) from the climate system. Removal times are of the order of 1000 years if carbon storage in deep-sea waters can be achieved, increasing to thousands to millions of years if burial in the sediments can be achieved.

With respect to ocean fertilization as a NET, there is specific interest in so-called High Nutrient Low Chlorophyll, or HNLC, regions. These contain sufficient amounts of major nutrients, while trace nutrients are in short supply—typically, the limiting trace nutrient is iron. However, not all ocean fertilization approaches are focused on trace nutrients such as iron. Some approaches instead focus on adding major nutrients, notably nitrogen and phosphorus, in regions where these are limiting.

Experiments with iron addition to HNLC regions have revealed that algal blooms can indeed be triggered, and that up to 10–20% of the triggered production of organic matter may sink to the sea floor under favorable conditions. One test reported that for each single iron atom, at least 13,000 carbon atoms were fixed into algal matter that derived almost exclusively from diatom algae and over half of the produced organic matter sank to depths below 1000 meters (Parry, 2012). But even the researchers behind that test were loath to extrapolate their results in terms of NETs capability. This is because of a healthy level of caution among scientists about manipulating a system we don't understand too well yet. The remainder of this section outlines key complexities associated with this approach (see also Table 4.1).

\* \* \*

First, it is critical to find out what type of algae will be stimulated most. Some algae calcify, which means that they build skeletal parts of calcium carbonate ($CaCO_3$), while others don't. Non-calcifying algae can build skeletal parts from silica ($SiO_2$) or can consist entirely of organic matter. There is a large difference in carbon management between non-calcifying and calcifying algae and thus in their impacts on the air–sea $CO_2$ exchange. Non-calcifying algae drive only the so-called organic-carbon pump, which results in drawdown of $CO_2$ in the surface ocean and thus the atmosphere, as just described.

**Table 4.1** Complexities associated with ocean iron fertilization (OF).

1. Uncertainties exist about which type of algae will be stimulated most
2. To avoid unbalanced ecosystems, OF will need to include balanced "cocktails" of other nutrients
3. Uncertainties exist about the efficacy of net carbon removal
4. OF may stimulate toxic or harmful algal blooms and/or harmful subsurface oxygen depletion.

Calcifying algae, however, drive not only the organic-carbon pump, which consumes $CO_2$, but also the calcium carbonate pump, which releases $CO_2$ into the environment according to the equation $Ca^{2+} + 2HCO_3^- \rightarrow CO_2 + H_2O + CaCO_3$. Non-calcifying algae therefore are much more efficient in driving $CO_2$ removal than calcifying algae.

There are indications that the global proportion of calcifying algae production will increase relative to that of non-calcifying algae in response to the combination of rising $CO_2$ levels from our unbridled emissions and increasing ocean stratification because of the attendant global warming (Rost et al., 2003). The associated gain in importance of the carbonate pump would lower the $CO_2$ uptake from the atmosphere by ocean plankton. This is not just some hypothetical scenario. All across the North Atlantic, this type of transition seems to be well under way, as indicated by an observed shift in the composition of algal remains in the sediment toward a reduced proportion of silica and an increased proportion of carbonate (Antia et al., 2001).

But there is a compensating mechanism as well. Rising $CO_2$ levels cause reduced calcification rates in key organisms such as corals, foraminifera, and coccolithophorids (Rost et al., 2003). This decreases the ratio of carbonate to organic matter production and thus increases the drawdown of $CO_2$ by marine organisms. The net outcome of the opposing effects of (a) the relative climate-induced increase in calcifying algae and (b) the $CO_2$-related decrease in biogenic calcification, forms a crucial feedback response to $CO_2$ increase that is not yet understood and that may work out either way.

* * *

Second, experiments have confirmed that iron fertilization can successfully trigger increased algal blooms and—applied in the right regions—can promote diatom productivity over that of other algae. Diatoms are non-calcifying algae that form skeletal parts from silica (opal), so that promoting diatom blooms avoids the drawbacks of the calcium carbonate pump. As a result, diatom blooms could be very effective at drawing down $CO_2$ in surface waters, and thus the atmosphere, transporting the carbon by means of dead organic matter to deeper waters and eventually the sea floor. Triggering diatom blooms in regions favored for fisheries or aquaculture has beneficial impacts because they are an excellent food stock. While that's great news, it doesn't make for a very effective NET. Being eaten by fish means that the carbon from the diatoms is digested and thus gradually released through the food chain, rather than permanently removed.

The performance of ocean fertilization as a NET is not easily assessed, largely because iron uptake by algae is not well understood. When detailed analyses were performed, they revealed some counterintuitive behavior. Notably, it was found that diatoms are greedy and use more iron than they need for photosynthesis; they then store the excess iron in their silica skeletons (Ingall et al., 2013). In plain speak, they are hogging the food, which may cause the environmental benefits of iron fertilization to be only short lived. Ingall et al. (2013) infer that iron-hogging by diatoms may lead to a long-term reduction in the amount of $CO_2$ that may be trapped by the ocean. Why? Because more iron fertilization might simply stimulate diatoms to take up even more iron for building even larger shells, rather than feed the growth of extra plankton. Upon death, the diatom shells sink to the ocean floor, and this removes precious iron from the surface ecosystem and starves off the diatoms' plankton peers.

The starvation concerns extend to other nutrients as well because production of biomass in the end depletes all resources. Lampitt et al. (2008) summarized the implications as follows: "Either these [resources] must all be subject to fertilization or [carbon removal] will eventually be limited by the one most slowly replenished . . . . Generally, phosphate is the ultimate limiting nutrient . . . , [and] would have to be included in the fertilization cocktail supplied."

\* \* \*

Third, iron fertilization may have only limited potential, and even large-scale implementation might remove only a fraction of human-caused emissions. This is because marine ecosystems are incredibly efficient at recycling organic matter. In most open ocean settings, on average, 1–25% of newly formed organic matter drops out of the surface layer. Of this, some 9 out of 10 parts are recycled in the deep pelagic and benthic ecosystems, and only 1 out of 10 parts is buried in seafloor sediments. So, on average, only 0.1–2.5% of organic matter produced by photosynthesis is buried away in the sediment. Crucially, only the proportion of organic matter that is buried in sediments represents a net loss of carbon from the climate system, and thus counts as a NET.

As a rule of thumb, a greater proportion of organic matter gets buried under highly productive, more wasteful, upwelling regions. Only below-average amounts get buried under nutrient-starved regions where ecosystems have evolved around being highly efficient at carbon recycling. Iron fertilization in HNLC regions with highly efficient recycling specifically aims to transform prevailing ecosystems, which have evolved over many millions of years toward efficient recycling, into wasteful ones that drive a high proportion of carbon loss to the sediments. By its very intent of achieving high carbon burial rates in sediments, therefore, the technique aims to effect drastic changes in the ecosystems. If such changes are implemented too rapidly, ecosystems cannot adapt. But nature has given us some examples of more gradual implementation of iron fertilization that did not cause ecosystem collapse and that we can learn from.

During ice ages, more dust was blown around in the atmosphere because of greater aridity and also because larger ocean shelf areas were exposed due to sea-level lowering. Iron contained in this dust caused changes in ocean ecosystems, which are often regarded as an important driver of increased carbon storage in ice-age oceans. Yet, it was not the only driver of $CO_2$ change. Precise picking apart of the various causes of carbon redistributions during ice-age cycles (Yu et al., 2016, 2019, 2020) reveals that the iron effect was assisted by lower temperatures that increased $CO_2$ solubility in water, increased sea-ice coverage that inhibited ocean outgassing, and changes in deep-water circulation that interacted with the biological organic carbon pump to favor more deep-water carbon storage. And here comes the critical bit: the ice-age carbon in the oceans was *not* buried in the sediments—it was held as dissolved inorganic carbon in oceanic deep waters. Remember that timescale of storage in deep-sea waters of the order of 1000 years? This means that new equilibria were reached during the ice ages at lower atmospheric $CO_2$ levels, with ocean uptake and release of carbon balancing each other out again. When the ice ages ended, the excess oceanic carbon was released again until new equilibria were reached at higher atmospheric $CO_2$ levels. The varying $CO_2$ levels during ice-age cycles operated as a critical positive feedback to the climate changes.

The natural ice-age experiments strongly suggest that iron-fertilization impacts on carbon removal and thus $CO_2$ reduction may be temporary (many centuries) rather than permanent. If nature could not manage to achieve permanent carbon removal through this mechanism during times of marked global dustiness that lasted thousands of years, then this raises some serious questions about humanity's ability to do so. From that perspective, one might argue that ocean fertilization doesn't stack up as a NET. But that would be too limited a view; after all, not everything needs to be picture perfect in all methods. Even if ocean fertilization does not achieve permanent carbon removal, its potential for carbon removal that lasts centuries may be just right. It would buy us time to get other, permanent removal approaches into full-scale operation. In other words, we should not discard any of the options on the table but adopt a flexible strategy toward a synergy that gets us to the desired end goal.

* * *

Fourth, iron fertilization may stimulate toxic or harmful algal blooms. Also, large-scale nutrient additions carry a serious risk of causing oxygen depletion as a result of decomposition (oxidation) of algal organic matter. Decomposition represents the opposite of the photosynthesis reaction, and therefore is represented by $CH_2O + O_2 \rightarrow CO_2 + H_2O$, which illustrates the oxygen consumption involved in this process. As oxygen levels drop in ocean waters, reagents other than oxygen become important. This risks production of the potent greenhouse gases $N_2O$ and methane, offsetting or overwhelming the initial ocean fertilization's reduction of greenhouse forcing through $CO_2$ removal. Finally, a large downward flux of organic matter in the ocean risks the development of anoxic (no oxygen) conditions, at least locally at the sea floor, and possibly more extensively throughout the water column. These are not just theoretical concerns; natural windblown-dust-driven ocean fertilization during ice ages left large stretches of the deep ocean considerably deficient in oxygen (Yamamoto et al., 2019).

Anoxic conditions are lethal to most life forms and thus would drive profound local to regional extinctions. Today, this can be observed already in the form of rapidly expanding anoxic ocean dead zones. Ironically, development of anoxia would be the single most favorable condition for promoting organic carbon burial in the sediments, exactly because anoxia would terminate most subsurface life in the ocean. The attendant collapse of the food web would mean that less organic matter is recycled, so that more remains available for burial. Yet, nobody in their right mind wants to create massive ocean dead zones. To achieve greatest efficacy with the least destruction, we need accomplish a delicate balance between creating a large downward flux of carbon in the ocean through fertilization, while not overwhelming the oxygen supply in the deep sea and at the sea floor.

Given the uncertainties of oceanic ecosystem response to ocean fertilization, its potential for carbon drawdown has been estimated over a wide range, with values varying from as little as 0.14 GtC per year to as much as 12 GtC per year.

## 4.3.2. Blue Carbon

The blue carbon approach extends beyond ocean fertilization in that it includes (re-)instating and protecting marshes, seagrass meadows, and mangrove and kelp forests in coastal regions (Figure 4.7). Again, the aim is to overcome recycling and achieve storage of excess carbon. The effort is pitched against large-scale die-off of these resources driven by climate change, such as the recent 95% Tasmanian kelp demise, part of a sharp global kelp decline (Bland, 2017). The kelp decline is echoed by a decline in mangroves, which is of particular concern because mangroves are among the most carbon-rich forests in the tropics. Organic-rich soils associated with mangroves account for half to nearly all of the carbon storage in these systems. Furthermore, coastal marine habitats are under severe, and often terminal, stress from pollution, eutrophication, and overfishing (e.g., Evans, 2020).

These coastal fringes have some of the richest carbon soils of all, and destruction of roughly half of them has resulted in a continuous stream of $CO_2$

**BLUE CARBON: THE $CO_2$ REMOVED
& STORED BY COASTAL ECOSYSTEMS**
Globally, these habitats are being lost at a rate of 1–2% a year

| MANGROVES | SALT MARSHES | SEAGRASS BEDS | KELP FORESTS |
|---|---|---|---|
| Extract around 30 million tonnes of carbon a year, provide critical habitats and absorb storm surges and rising tides. | Once cleared for grazing land, many are now being restored to provide valuable coastal defences. | These flowering, submerged plants are extremely carbon-rich, but also vulnerable to river pollution. | Giant seaweed forests found in cold waters are being affected by rising sea levels and ocean warming. |

**Figure 4.7**  The dominant blue carbon environments.

*Source*: Based on Pearce (2019); Blue Carbon Initiative; Sanderman et al (2018). https://chinadialogueocean.net/11915-coastal-ecosystem-natural-solutions-climate-change/

emissions into the atmosphere. The emitted carbon had been stored for centuries and therefore is commonly counted as so-called external carbon that was locked away from the climate system on a long-term basis. In other words, deteriorating coastal systems count as sources of greenhouse gas emissions. Following decades of deleterious actions against mangroves and other coastal ecosystems, they now emit 0.04–0.28 GtC per year, which is equivalent to 3–19% of the emissions from global deforestation.

Coastal system regeneration is clearly pivotal for avoiding further emissions from deteriorated coastal systems, which would exacerbate climate change, as well as returning the carbon already emitted back into these systems. Regeneration of these systems can put considerable amounts of carbon back into the ground in high densities and at relatively low cost. And there are significant associated benefits to fisheries, coastal environmental and hazard protection, biodiversity, and water quality. Application of electrically stimulated limestone growth (Biorock; see section 6.3) can substantially increase the growth of marine plants where other methods fail, and thus make significant contributions to carbon storage and nitrogen fixation. Preserved and enhanced blue-carbon systems have an especially important benefit in that they greatly improve coastal resilience against the wider impacts of climate change, such as sea-level rise, storm-driven erosion, and storm-surge inundation.

All things considered, coastal blue carbon initiatives fall more under the categories of emissions reduction and emissions avoidance than under NETs. Going forward, the main hope is to undo the damage we have inflicted on these systems, but there is little hope that we will be able to turn the tide and achieve growth beyond what existed in pre-industrial times. Thus, the contributions of blue carbon to net carbon drawdown are limited at best, but blue carbon initiatives remain very important because of all their associated benefits. A key issue that needs to be resolved for coastal blue carbon implementation concerns potential land-use competition with initiatives for property and coastal-defense-structure development.

## 4.3.3. Ocean Alkalinization and Negative Emissions Hydrogen

Alkalinity is the opposite of acidity. Hence, ocean de-acidification is technically known as ocean alkalinization. On a scale from acidity to alkalinity, natural ocean water is slightly alkaline. Ocean acidification is pushing it toward lower and lower alkalinity, which reduces the water's capacity to absorb $CO_2$ from the atmosphere. Ocean alkalinization aims to drive conditions back the other way, which will increase the water's capacity to absorb $CO_2$.

Several ocean alkalinization techniques have been proposed (Renforth and Henderson, 2017) (Figure 4.8). They typically involve the addition of carbonate or silicate minerals in seawater to react with dissolved inorganic carbon in the ocean, though some applications are combined with land-based enhanced weathering, with run-off from the land through rivers carrying the dissolved mineral ions into the ocean. Either way, alkalinization of ocean water reduces dissolved $CO_2$ and thus causes the ocean to absorb more $CO_2$ from the atmosphere.

One of the main ocean alkalinization NETs that have been proposed is tidal/wave-zone weathering of olivine "green sands" that would be spread over beaches. This approach utilizes wave energy to continually grind olivine particles to expose fresh mineral surfaces for weathering. Another ocean

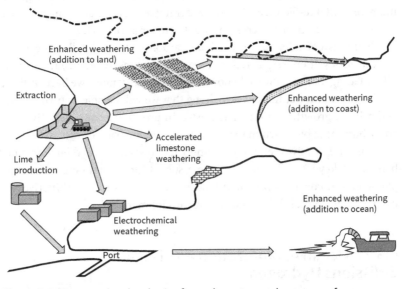

**Figure 4.8** Proposed technologies for carbon storage by means of ocean alkalinization.

*Source*: Based on Renforth and Henderson (2017). https://agupubs.onlinelibrary.wiley.com/doi/full/10.1002/2016RG000533

alkalinization approach is ocean liming, which uses dispersal of dissolved lime (calcium oxide and calcium hydroxide) into sea waters to convert dissolved $CO_2$ into calcium carbonate. And a further major proposed method is direct-buffering mineralization of limestone or silicates with $CO_2$-enriched water under carefully managed conditions in coastal water-treatment plants. Finally, a novel suggestion concerns negative emissions hydrogen, which centers around sea-water electrolysis and feeding managed waste waters back into the ocean.

Elements of mining, grinding, reaction-temperature management, and dispersal in these techniques require energy. Hence, total life-cycle greenhouse-gas assessments are essential to ensure that methods represent true NETs. The use of low-cost renewable energy therefore forms a critical requirement for effective deployment. In the case of ocean liming, $CO_2$ is produced during lime production both by the reaction itself and as a by-product of the energy used for running the kilns as well as during the broad shipping-based dispersal that is needed to avoid over-saturating the seas at input sites, with harmful side effects. To enhance the efficacy of ocean liming as a NET, these $CO_2$ emissions must be managed carefully by $CO_2$ capture at kilns and de-carbonized transport options.

Negative emissions hydrogen, also referred to as electrochemistry, is a new marine NET concept (Rau et al., 2018; The Climate Examiner, 2018). At its heart lies the production of hydrogen through electrolysis of saline sea water. By itself, that does not constitute a NET, and it would even be a source of emissions if the energy for electrolysis is obtained from fossil fuels. In consequence, two measures are taken. First, energy will be used from renewable sources to reduce emissions. Second, NET potential is created by addition of carbonate or silicate minerals—often available as waste products from mining and industry—to the wastewater from electrolysis. This creates alkaline waste brines that are highly reactive with $CO_2$. Carefully dispersed in the oceans, such brines reduce dissolved $CO_2$, alleviating ocean acidification, and air-sea exchange then drives commensurate lowering of atmospheric $CO_2$ levels. This way, negative emissions hydrogen can produce both hydrogen and substantial negative emissions potential.

The green-sands approach does not fall under the strictures of the London Protocol, but ocean liming likely will. Ocean alkalinization as part of negative emissions hydrogen might escape the London Protocol if the disposal of alkaline seawater could be considered incidental to the usual operation of hydrogen production. This opens up considerable perspectives for these NETs, in contrast to ocean fertilization.

There are some issues around iron fertilization associated with the green sands approach. This is because the minerals used contain iron that, once liberated, may drive regional increases in algal productivity. Unwitting enhancement of coastal dead zone development then is a matter of concern.

Development of ocean alkalinization approaches is in its infancy. But early studies suggest that the potential for carbon drawdown lies somewhere between 0.27 and 2.7 GtC per year. Concepts for negative emissions hydrogen are even less maturely developed, and estimates for potential carbon drawdown are in the typically optimistic phase of early excitement: baseline values from 25 GtC per year have been suggested for full implementation (Rau et al., 2018). My synthesis uses a careful range of 10–50% of that value, but deeper research is needed to constrain the range to less arbitrary values.

## 4.4. Land-based Technological NETs

Technology-based NETs on land revolve around different means of capturing $CO_2$ directly from the atmosphere and then processing it for long-term storage. Some of the capturing methods involve vegetation growth. Others are of a manipulated, chemical nature. In our discussion of NETs, $CO_2$ capture alone is not enough; it must be coupled with long-term storage options for the captured $CO_2$. And any emissions across the entire processing and supply chain must be less than the amount of $CO_2$ removed, so that a true net removal takes place. The key technology-based NETs on land are Bio-Energy with Carbon Capture and Storage, or BECCS, and Direct Air Capture with Carbon Storage, or DACCS. The previously discussed PyCCS (section 4.2.2) is being mooted as an alternative to BECCS, or something that could be applied under different circumstances.

## 4.4.1. BECCS and PyCCS

Bio-Energy with Carbon Capture and Storage has rapidly gained prominence among carbon removal approaches in the climate change debate, especially because it was adopted by IPCC in their modeling scenarios (Hickman,

2016). In BECCS, biomass is used instead of fossil fuel to generate energy. The biomass that is used can derive from forestry and crop waste, or from specially planted forests and crops. The CCS aspect relies on either the addition of $CO_2$ capturing systems to existing energy facilities or the deployment of new facilities with such systems built in from the outset. The captured and concentrated $CO_2$ can then be stored through geological processes or injection into reservoirs (section 4.2.3) or through conversion into mineralized carbon products (section 4.2.4) (Figure 4.9).

When used with efficient CCS, and with careful emissions management during the growing, harvesting, and transporting of the biomass feedstock supply, BECCS can result in a net emissions result that is truly carbon neutral or even carbon negative. But this outcome is by no means a given; very careful full life-cycle assessment is needed to ensure that all processes are optimized in such a way that this outcome is achieved (Tanzer and Ramirez, 2019). Regardless, BECCS offers the potential of a reliable source of power, while at the same time creating revenue streams from increased demand for forestry output, capturing value from logging and agricultural waste. Moreover, there is potential for linking BECCS with waste from carbon-negative regenerative agriculture.

The storage component of BECCS needs to ensure that $CO_2$ cannot return into the climate system. Such return may happen, for example, (a) if the $CO_2$ is used in synthetic fuels or fizzy drinks and other temporary substances,

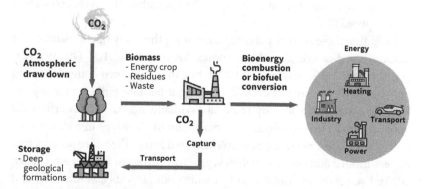

**Figure 4.9**  Concept schematic of BECCS.

*Source*: Following Consoli (2019). https://www.globalccsinstitute.com/wp-content/uploads/2019/03/BECCS-Perspective_FINAL_18-March.pdf

(b) when there is storage-reservoir leakage, (c) if $CO_2$ storage is used for enhanced recovery of hydrocarbons (which lead to new emissions) from the reservoir, or (d) because of potential deep biosphere processes. When everything is optimized, global upscaling of BECCS can result in estimated carbon drawdown between about 0.13 and 3.3 GtC per year (Smith et al., 2016b; Fuss et al., 2018). However, there are issues with the higher-end estimate in particular. The limiting factor for BECCS is the supply of biomass. To achieve the high estimate, massive dedicated forestry would be needed, which would compete heavily with food production for land surface area and water and nutrient resources. It has been estimated that an area of arable land equivalent to twice the size of India would be needed and that the specific biomass increase would be equivalent to three times the world's total cereal production, requiring twice the annual global agricultural water use and several times the annual nutrient use (Consoli, 2019 and references therein).

Given that competition with resources for food production is a serious concern with BECCS, it is tempting to shift focus for growing BECCS feedstock onto forested land or ancient grassy biomes. However, as we have seen in section 4.2.1, harvesting an ancient forest and replacing it with plantations will reduce the carbon stock in both trees—at least for a period of 50 to 120 years until the replanted trees reach the same level of carbon storage as the originals—and soil. And developing plantations on ancient grassy biomes may cause substantial soil carbon release as well. These emissions from land-use change might be avoided by leaving the original systems alone. They need to be considered carefully in the full carbon life-cycle assessment of a proposed BECCS plant.

Including these further potential emissions, the likely upper estimate for BECCS-based carbon drawdown comes down to about 1.35 GtC per year (Williamson, 2016; Fuss et al., 2018), while the drawdown at the lower end may be 0 GtC per year or even less than that because emissions along the lifecycle compensate for any removal (Williamson, 2016). Powerful calls have been made to de-prioritize bioenergy and BECCS-type installations because the carbon drawdown potential assessed across the entire process may be very limited and because such developments threaten wild forests (Baker, 2020). Even if the net carbon budget works out favorably, biodiversity and ecosystem services are likely to suffer significantly. Relying on BECCS alone to keep warming within the Paris Agreement's 2°C limit could thus do more damage to terrestrial ecosystems than 2.8°C global warming would

(Williamson, 2016). A fundamental rethink of this much-discussed process seems to be required.

A new concept aims to resolve the problem that widespread plantations for BECCS would replace and thus reduce agricultural food-protein production. It combines BECCS with algal production, giving rise to the abbreviation ABECCS (Beal et al., 2018). In practical terms, a major culturing facility for algae would be co-located with a BECCS facility. The BECCS facility then provides both (bio-)energy and $CO_2$ to the algal facility, where algal growth removes the $CO_2$. Remaining $CO_2$ would go to long-term storage, as in normal BECCS. The cultured algal protein would provide feed ingredients for poultry, swine, sheep, cattle, seafood, and humans. This compensates for the loss of food-protein production related to replacing agricultural crops by plantations for the BECCS facility. While the $CO_2$ component used in algal culturing recycles back into the climate system upon consumption and decomposition on short timescales, and therefore does not constitute a NET component, any long-term storage of excess $CO_2$ still adds a true NET component. The study indicates that some net $CO_2$ removal can be achieved with such a set-up.

Schmidt et al. (2018) raise another issue that is worthy of deeper investigation: when evaluating biomass production for its NETs potential, the key factor in determining the land-surface area required for biomass production is the so-called carbon sequestration efficiency (CSE). Increasing the CSE decreases the land requirement. They then highlight that proposed BECCS-based NETs have a theoretical CSE of 90%, which can only be attained at high economic costs and environmental risks. PyCCS systems (section 4.2.2; Figure 4.3) can attain a similar theoretical CSE when $CO_2$ capture and storage is included, or more realistic values of 70–80% when only biochar and bio-oil are sequestered.

On this basis, the study infers that, instead of establishing vast, problematic industrial biomass plantations to feed BECCS, "a network of smaller-scale PyCCS systems ... [might] establish biomass as a desired and cultivated coproduct of agriculture, for example in agroforestry, silvopasture, and perennial coppicing types of agriculture (i.e., carbon farming approaches ... )." Applied on a sufficient scale in a widely distributed manner, PyCCS provides a carbon drawdown potential of about 1.4 GtC per year, similar to the high-end estimate for BECCS. On the downside, the study acknowledges that the logistics of such farm-based

PyCCS systems, producing biochar for local soil applications and bio-oil for centralized processing facilities, may be challenging for nations with limited infrastructure and organization. Yet, similar challenges may be foreseen for large-scale BECCS.

## 4.4.2. DACCS

Direct Air Capture, or DAC, in essence consists of massively upscaled versions of the concept of $CO_2$-scrubbing from air that has been applied for more than 60 years in spacecraft and submarines (Li et al., 2013). Those were developed because build-up of $CO_2$ in closed-off spaces causes a range of problems that can become very serious (Table 4.2). The scrubbers traditionally used a strong chemical base—soda lime, ammonia-derivatives called amines, or lightweight lithium hydroxide in space—to absorb the $CO_2$. In a next stage, the reaction products are processed to separate the $CO_2$ again in pure form. Later versions have been built around regenerative metal oxides, molecular sieves and membranes, and metal-organic frameworks. Once captured, the $CO_2$ can be ejected overboard or stored under pressure for later use or disposal.

**Table 4.2** Influences of $CO_2$ concentrations on humans.

| Concentration range (ppm) | Effects |
| --- | --- |
| 350–1000 | No issues: typical level in occupied spaces with good air exchange |
| 1000–2000 | Poor air: drowsiness |
| 2000–5000 | Stagnant, stale, stuffy air: headaches, sleepiness, loss of concentration and attention, raised heart rate, nausea; 5000 ppm is the permissible exposure limit |
| above 5000 | Toxic and/or low-oxygen air: oxygen deprivation may occur |
| above 10,000 | Dangerous oxygen deprivation: convulsions, coma, and death |
| above 30,000 or 40,000 | Loss of consciousness in seconds, leading to death |

From: Bonino (2016); Permentier et al. (2017)

$CO_2$ levels in the climate system fall well below the range that is toxic to humans. That is not the issue of concern. For climate change, the issue of concern is the role of $CO_2$ in the radiative balance of climate, and it is dominated by more subtle changes in the atmospheric $CO_2$ concentration (Chapter 2). For climate, therefore, different types of $CO_2$ scrubbers are needed. They need to function efficiently at low concentrations of 500–300 ppm, they need to deal with very large quantities of air, and they need to be suitable for massive upscaling if they are to provide meaningful assistance with humanity's Gigatons challenge. These are not trivial requirements.

Current DAC demonstration facilities use liquid lime-based solvents or solid amine-based sorbents to trap $CO_2$ from air that's fed through the system. Thereafter, the system is flushed with heat, in some cases in a vacuum, which causes the $CO_2$ to be released, and it is then captured in concentrated form. Both approaches have high energy demands for air handling and for flushing the system with heat and/or to maintain a vacuum. New approaches with regenerative metal oxides, molecular sieves and membranes, and metal-organic frameworks are being developed, but are not yet implemented in industrial-scale test sites.

DAC could be our most environmentally friendly means of large-scale $CO_2$ removal, especially when driven with renewable energy (Williamson, 2016). But DAC alone is not a NET. As with BECCS, the method only becomes a NET through coupling with high-volume, long-term $CO_2$ storage. DAC then becomes known as Direct Air Capture with Carbon Storage, or DACCS (Figure 4.10). Suitably upscaled, DACCS could be a significant player in carbon dioxide removal, but—like all NETs, or even all NETs together—not to the extent that their possible future application gives any excuse to postpone drastic emissions reductions (Gambhir and Tavoni, 2019).

Currently, one facility delivers $CO_2$ for synthetic fuel production, another for increasing vegetable production in greenhouses, and a third for carbonation of soft drinks. Captured $CO_2$ from these applications returns into the climate system on relatively short timescales. In consequence, these facilities demonstrate the feasibility of DAC, but do not qualify as NETs.

Only the Climeworks CarbFix2 pilot plant, which is co-located with a Reykjavík Energy geothermal powerplant in Iceland, advances the concept to DACCS, which is a true NET. It employs $CO_2$ injection into basalts and runs on renewable energy. But, so far, the longevity of $CO_2$ storage by injection into basalts is not yet fully understood because of apparent interactions

**Figure 4.10**  Conceptual representation of DACCS, with $CO_2$ stripped from the open atmosphere and stored in a geological reservoir.

*Source*: Based on Appel (2019). https://towardsdatascience.com/
tackling-climate-change-with-machine-learning-78d1e185b3ec

with the deep biosphere (section 4.2.3). Questions also remain with respect to potential and longevity of geological storage options. These include poorly understood interactions with microbial communities (Geospace, 2011) and with the reservoir fluids themselves (Cohen and Rothman, 2015), as well as potential leakage at injection wells and through permeability and cracking of cap-rock formations (Breyer et al., 2019). Safer storage may require conversion of the $CO_2$ into solid materials, but this is difficult to upscale into the range of Gigatons of carbon. It also would require further energy expenditure and thus potential emissions.

In addition, the technology is not nearly at the scale required. As yet, the three listed DAC facilities and the one DACCS facility trap annual $CO_2$ quantities of about 400 tons (now being upscaled to 1 million tons), 1000 tons, up to 50,000 tons, and about 12,000 tons, respectively. To calculate the corresponding tons of carbon, divide these numbers by 3.67. Even when there is storage, question marks remain about its longevity, which require substantial further research before proper full lifecycle emissions assessments can be made to determine the true NET potential.

It is great news that industrial-scale test facilities are now operational and that $CO_2$ recycling is possible to at least promote avoidance of new emissions. And, given enough investment, there is little doubt that the DAC component can be scaled up to scales of Gigatons of carbon capture. But to address humanity's Gigatons challenge, coupling is needed with long-term $CO_2$ storage options at sufficient scales and in ways that avoid its return into the climate system. Key options are through geological processes or injection into reservoirs, or through conversion into mineralized carbon products. Emissions likely occur during processing, transport, and storage. Like BECCS facilities, therefore, also DACCS plants need meticulous full-lifecycle emissions assessment and monitoring, to determine the true extent to which net carbon removal is realized.

Wider impacts of NETs include land surface area requirements, social acceptability, and water requirements. DAC needs only limited land surface area relative to other NETs, and there will not likely be a fundamental clash with space requirements for food production. To address humanity's Gigatons challenge, however, many of these large-scale industrial facilities will be needed, and all will require substantial amounts of renewable energy. Such widespread, highly visible industrial development may cause friction among communities. With respect to water, Smith et al. (2016b) summarize that DAC using sodium hydroxide has relatively modest requirements of 3.7 tons of water per ton carbon. Using amines, as in Climeworks' system, the same study suggests theoretical requirements of about 92 tons of water of water per ton carbon. However, Climeworks (2020) states from their practical experience that their plant does not need external water supply because it extracts water directly from the vapor in the air that is processed. It remains to be established whether, and to what extent, this direct water recovery would work outside humid regions like Iceland, for example in arid subtropical regions, but it suggests at least that theoretical studies may have overestimated the water requirements.

Finally, DACCS costs are often reported as prohibitive. A recent assessment of the DAC component, driven by renewable energy, indicates costs per ton of $CO_2$ captured, or 0.27 tons of carbon captured, of $120, $80, and $60 by 2030, 2040, and 2050, respectively. This might be dropped even further to

$35 per ton of $CO_2$ under a best-case scenario in North West Africa, which is replete with renewable energy potential from solar PV (Breyer et al., 2019). It needs to be emphasized that these cost estimates are considerably lower than those from other assessments, which typically are in excess of $100 per ton of $CO_2$ (e.g., Henderson et al., 2018; National Academies of Sciences, Engineering, and Medicine, 2019).

Adding carbon storage methods, and accounting for any emissions during the entire process, raises these costs, just like it would with BECCS, and with emissions-reduction approaches that involve storage of $CO_2$ captured at in-dustrial point-sources of emissions. There's no denying that DACCS is among the more expensive NETs in terms of investments needed. Regardless, it may be indispensable because of the sheer volume of $CO_2$ removal that might be possible and because of its relatively modest wider environmental impacts.

In sum, we're still a considerable way from running direct air capture as a true NET at sufficient scale to address humanity's Gigatons challenge. But other ways of carbon removal are ready to be implemented at larger scale, such as reforestation/afforestation, soil enhancement and regeneration, and enhanced weathering. This provides a window of time for DACCS to be perfected and join the portfolio at scale, taking the carbon removal to great heights in later decades. Such a phased implementation approach will be dis-cussed more extensively in section 4.6 and Chapter 8, as a so-called NETs portfolio approach.

## 4.5. Marine Technological NETs—Artificial Upwelling

Artificial upwelling is designed to trigger algal blooms by bringing colder, more nutrient-rich waters from depth up to the surface. Generally, the same caveats apply as for ocean fertilization (Figure 4.6; section 4.3.1). The deeper waters can be brought to the surface using giant pumps, or—more elegantly—using giant vertical plastic pipes in the ocean, tethered between the sea floor and a buoy at the sea surface (Figure 4.11). These pipes have flap valves that allow water to enter only from below and exit only at the top; ocean wave action provides the pumping action. The tubes are designed to be 200–1000 meters long and 1–10 meters wide. To achieve an increase in

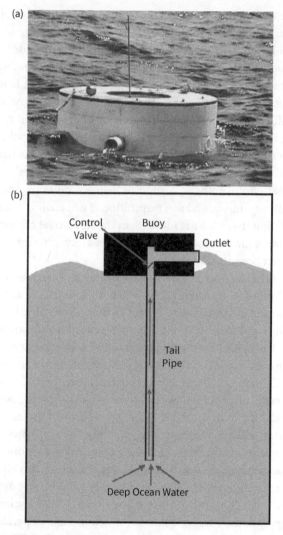

**Figure 4.11** Artificial upwelling using a vertical pipe. Left: Buoy with water outlet. Right: Schematic of the system's operation.

*Source*: Zhang et al., (2016); http://www.wrrc.hawaii.edu/bulletins/2000_12/aumix.html. Used by permission, ScienceDirect.

net oceanic carbon uptake of 1 GtC per year, several millions of such tubes would be required (Yool et al., 2009).

Yool et al. (2009) consider just the ocean, but a broader perspective is warranted. When considering the entire Earth System, Oschlies et al. (2010) found a reduction in atmospheric $CO_2$ levels that is much larger than the uptake of carbon into the ocean. Artificial upwelling brings not only nutrients but also cooler water to the surface. In consequence, widely applied artificial upwelling causes sea-surface cooling. This, in turn, causes lowering of surface air temperatures and consequently of continental soil temperatures. This cooling has two effects: it drives a decrease in net photosynthetic carbon drawdown on land, and it causes a reduction in vegetation and soil respiration. The second effect outweighs the first, so that there is a net increase in the terrestrial carbon inventory, which explains the disproportionate drop in atmospheric $CO_2$ relative to the minor net oceanic $CO_2$ uptake.

So, why is the net oceanic $CO_2$ so small for artificial upwelling? There are two contributing factors here. First, apart from nutrients and lower temperatures, artificial upwelling brings water enriched in dissolved $CO_2$ to the surface. This is the downside of the higher nutrient concentrations in deep water. Those originate from decomposition of sinking organic matter, and the same process releases $CO_2$. This is why natural upwelling is normally a source of $CO_2$ to the atmosphere, and artificial upwelling mimics this, except in some specific places of the world's oceans (Oschlies et al., 2010). Second, the loss of $CO_2$ from the ocean caused by productivity increase in response to artificial upwelling is for a very large part offset by rapid decomposition of the same material. This efficient recycling is something we discussed before under ocean fertilization (section 4.3.1). As a result of this efficient recycling, even a large increase in algal productivity would result only in a small increase in net $CO_2$ drawdown from the ocean. This recycling argument does not apply to the cooling influence, which therefore operates at a much greater efficiency. As a result, the cooling influence on the land-based carbon cycle is set up to dominate over the net oceanic $CO_2$ change, which is exactly what the study found.

At least, one might say, artificial upwelling has a good chance of drawing down $CO_2$ levels. In the Earth System study, the hypothetical most optimistic assumption of a massive deployment of perfect ocean pipes achieved a removal rate of 0.9 GtC per year. At the same time, global temperature lowering of about 1°C is achieved in that most optimistic scenario. Lower surface ocean temperatures in the tropics moreover would likely reduce the frequency and/or intensity of tropical storms. These are all generally desirable prospects, so that more research seems warranted. Importantly, this does not have to remain limited to modeling studies, given that observational research has become possible since two pump-based experiments became operational, offshore at Hawaii and in the East China Sea (Geoengineering Technological Briefing, 2018b). However, it will be very difficult or virtually impossible to establish the exact quantity of net $CO_2$ removal by this method, primarily because the $CO_2$ changes are importantly focused on the land-based carbon cycle, far away from the artificial upwelling sites.

As always, there are downsides. A first downside is that the massive enhancement of natural productivity required to achieve globally significant net oceanic $CO_2$ uptake opens the prospect of important ecological and deoxygenation impacts from artificial upwelling, as with any ocean fertilization scheme. But with artificial upwelling, the relatively modest net oceanic $CO_2$ uptake may at least be augmented by—possibly larger—net $CO_2$ on land, in a way that does not occur with regular ocean fertilization.

Another downside is the massive implementation scale needed to make a substantial impression on $CO_2$ levels, which involves installation of millions of pipes of very substantial dimensions and mass in the open ocean. This will create issues for shipping, fisheries, and potentially wildlife, although the latter is usually quite able to adapt to and even profit from man-made structures. Also, the ocean is a hostile, highly energetic environment; damage to and loss of equipment must be expected, similar to what happens with a lot of oceanographic equipment. When massive ocean pipes go adrift in an unpredictable manner, heavy damage to other equipment and ships becomes a very real risk. The reality of this risk is well illustrated on a smaller scale by rogue shipping containers, for example.

There are also concerns about the fact that the required millions of pipes are made of plastic, which will over time—especially when damaged or old and weathered to bits—add yet more harmful particle pollution to the oceans. Finally, studies indicate that termination of artificial upwelling would lead to shock warming (Oschlies et al., 2010) and potential harmful algal blooms (Geoengineering Technological Briefing, 2018b).

## 4.6.  Synthesis

Taking an overview of the various NETs discussed, it is evident that we will need a broad portfolio of approaches. This is crucial to cover for three issues. First is the fact that none of the NETs by itself will, any time soon, be potent enough to meet the full Gigatons challenge. This can be assessed by looking at the minimum and maximum expected carbon removal across a range of studies. Second, NETs exist in different stages of readiness for deployment, so that staged implementation seems a natural course of action. Third, it is simply imprudent to put all our eggs in one basket, as any theoretically promising method, or even one promising in small-scale trials, may turn out to be a dud at full scale in real life. For example, this could happen because a method might be demanding of much higher energy input than considered before, or because the carbon storage component proves to be much more inefficient than thought before. The second and third issues are commonly evaluated using the so-called nine-stage Technological Readiness Level (TRL) scale (EC, 2014), which was based on an original concept developed by NASA in the 1970s (Table 4.3). TRL 1 is the level where only the underlying basic principles are observed, and many unexpected surprises may be lurking around the corner. TRL 9 is the level assigned when the actual system is observed to be working (at scale) in the operational environment. These are the tested and tried methods, with limited chances of unpleasant surprises. Other

**Table 4.3** Technological Readiness Levels (TRLs).

| Level | Description |
| --- | --- |
| TRL 1 | Basic principle observed |
| TRL 2 | Technology concept formulated |
| TRL 3 | Experimental proof of concept |
| TRL 4 | Technology validated in laboratory (or small field trial) |
| TRL 5 | Technology validated in relevant environment |
| TRL 6 | Technology demonstrated in relevant environment |
| TRL 7 | System prototype demonstration in operational environment |
| TRL 8 | System complete and qualified |
| TRL 9 | Actual system proven in operational environment |

considerations with respect to NETs are the requirements for fresh water, and land-surface area, given that these set a method up for potential competition with food production for these resources.

In Figure 4.12, I have brought together information from a range of studies and reports on the potential GtC capture per year (minimum or likely value and maximum), TRL, environmental impact scores, freshwater requirements, and land area requirements. Note that this information is shown for illustrative purposes only and that it cannot be considered a binding assessment without independent validation against the indicated sources.

My impact scores in Figure 4.12 have been assigned between 0 = safe and 1 = highly detrimental. They are subjectively based on general considerations, for example regarding ocean anoxia and ecosystem changes under ocean fertilization and artificial upwelling; biodiversity and native forest impacts under BECCS implementation; competition for water, land surface, and nutrient requirements with other sectors such as food production; anticipated issues with regional release of alkaline solutions from ocean alkalinization of negative-emissions hydrogen applications; environmental impacts of required infrastructure; potential negative consequences such as albedo change; and potential conflict with requirements of resources other than land and water for other uses. Table 4.4 lists the arguments I have taken into consideration when assigning my subjective impact ratings.

In Figure 4.13, I present an index score for the various NETs based on the information shown in Figure 4.12. This index incorporates the carbon removal potential, the TRL, and land, water, and other environmental impacts. High scores on the index are favorable; low scores suggest a poor relative balance between carbon removal and developmental stage versus the wider impacts. In other words, high scores are of particular interest, while low scores require very serious thought about the process. This thought should then especially focus on hard measures to alleviate impacts, along with increasing the TRL. Interestingly, the much-reported BECCS falls into the category of needing very serious further thought because of its great resource requirements and potential impacts (Figure 4.13). Soil carbon storage (SCS) seems the more ready-to-go solution with considerable carbon drawdown potential, especially when combined with biochar (or PyCCS) and enhanced weathering.

The use of TRL in my index for ranking methods relative to one another requires a bit of attention, as was pointed out to me by a friend, policy and strategy

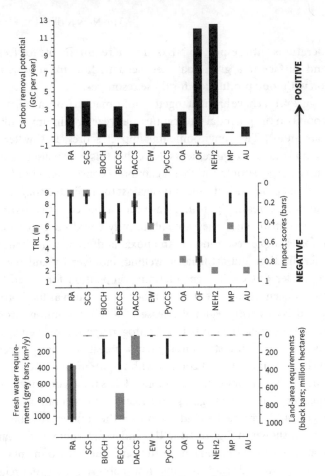

**Figure 4.12** Synthesis for the various NETs. Top: minimum to maximum carbon removal potential. Middle: TRL and rough minimum to maximum environmental impact score between 0 and 1 (see Table 4.4). Bottom: minimum to maximum requirements for fresh water and land surface area.

RA: reforestation and afforestation; SCS: soil carbon storage; BIOCH: biochar; BECCS: bio-energy with carbon capture and storage; DACCS: direct air capture and carbon storage; EW: enhanced weathering; PyCCS: pyrogenic carbon capture and storage; OA: ocean alkalinization; OF: ocean fertilization; NEH2: negative-emissions hydrogen; MP: mineralized products; and AU: artificial upwelling. Also indicated is the direction of impact for the various attributes: positive means best capture, highest TRL, and least environmental impact, and vice versa.

*Sources*: This diagram synthesizes results from Shepherd et al. (2009), Yool et al. (2009), Oschlies et al. (2010), Gadikota et al. (2015), Smith et al. (2016a,b), Williamson (2016), Fuss et al. (2018), Henderson et al. (2018), Lawrence et al. (2018); Minx et al. (2018), Rau et al. (2018), Schmidt et al. (2018), Bastin et al. (2019), Levin (2019), IPCC (2019), National Academies of Sciences, Engineering, and Medicine (2019), Paustian et al. (2019), and Bossio et al. (2020).

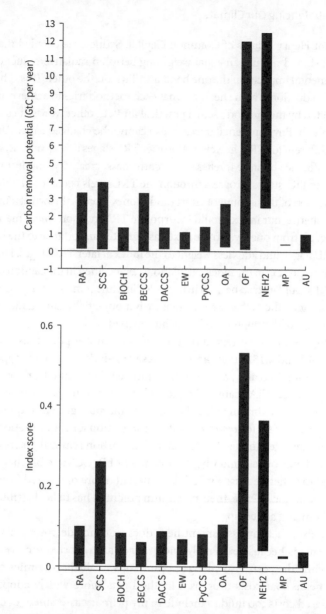

**Figure 4.13** Overall evaluation for NETs. Top: Minimum to maximum carbon removal potential, as in Figure 4.12. Bottom: Index scores as a tentative indication of feasibility and/or desirability range. The higher the value, the more feasible and/or desirable a method appears to be based on the attributes that are included. Underpinning values, arguments, and method for the index scores are elaborated in Table 4.4.

consultant Henry Adams of Common Capital, Sydney. He detailed that such an index should not use an equal weighting between radiative reduction and environmental impacts on the one hand and TRL on the other. This is because radiative reduction is at the heart of what each method tries to achieve and will be restricted by the method's actual practical limit; in other words, it cannot be moved much. Environmental impacts also cannot be changed easily. The situation is different for TRL, however. At earlier TRL stages (lower values), it is less easy to influence because it relies on inventiveness, creativity, and innovation. But at later TRL stages, progress through the TRL levels becomes a straightforward function of the amount of effort (and money) that is thrown at a method. In consequence, our index should incorporate TRL in a non-linear manner, so that progress from one level to the next has a greater importance to the index at early TRL stages, and adds less weight to the index at later TRL stages. I have applied this reasoning in my index calculations, as specified in the caption of Table 4.4. If TRL would instead be used in the index in a linear manner, then we would risk favoring methods that are very mature but potentially quite dangerous, to the detriment of development of better alternatives.

An important caveat regarding the carbon removal potentials shown in Figures 4.12 and 4.13 is that these are these are all determined independent of one another. In reality, however, they interact. For example, reforestation and afforestation (RA) cannot be pushed to maximum values if BECCS is also pushed to maximum values because the biomass growth requirements for BECCS compete for resources with reforestation and afforestation. Also, if RA is undertaken to fuel BECCS, then their carbon removal potentials can be counted only once. Similarly, the potentials of PyCCS and biochar cannot be added together because PyCCS is a different, more optimized form of biochar production. Hence, their maximum potential has to be distributed between them, not added up.

Note, also, that my assessment here does not include costs per GtC removal. This is because extremely wide ranges are reported for the various NETs, while the costs also change because of so-called economies of scale, which means that costs drop as a method gets more widely adopted. So, I considered costs too fluid to include in my primary assessment of capacity and technical and environmental feasibility. However, in Chapter 8 I do illustrate the implications of including today's cost estimates, which demonstrates that including cost considerations does not really change the picture by much.

Assessment of the index scores (Figure 4.13), in conjunction with the information behind the index scores (Figure 4.12; Table 4.4), offers a degree of objective insight into prioritization of NETs, and on the specific adjustments that could bring lower-rated ones into a higher priority ranking. SCS clearly comes out as a prime NET to be pursued in the short term because it has a high maximum index score while soil carbon restoration also brings important added benefits, such as improved water retention and productivity. Although very new and relatively poorly understood, negative-emissions hydrogen (NEH2) comes out as a method that warrants serious attention because of its strong potential. Ocean fertilization (OF) similarly shows considerable promise, if its large uncertainties with respect to longevity of carbon removal and international legal agreements can be overcome. Overall, NEH2 and OF score high because of the ocean's massive potential capacity for carbon removal, although it remains to be demonstrated that this potential actually is accessible.

The index suggests that Mineralized Products (MP) production is a sound approach for ramping up to worldwide scales; although its total carbon capture potential may be on the small side (Mac Dowell et al., 2017), there are significant other benefits such as the usefulness of the products themselves. DACCS, enhanced weathering (EW), ocean alkalinization (OA), and biochar and/or PyCCS also show substantial promise. For these, development is still needed that focuses specifically on excluding the possibility of limited returns, or—in other words—to further optimize the long-term carbon drawdown potential. All things considered, RA and BECCS present surprisingly low index scores. This is mainly because of their great risk of major environmental impacts, specifically with respect to their water and land surface area requirements, and also regarding detrimental impacts on biodiversity and wider ecosystem services. These measures could be turned into more promising ones if/when solutions are found for those drawbacks. Artificial Upwelling (AU) is still fraught with issues and at a very immature stage of development. It would require much development before it can be seriously considered.

Besides those details about individual NETs, it is abundantly clear from the carbon removal potentials shown in Figures 4.12 and 4.13 that a wide-ranging portfolio of NETs implementation will be needed, with phased implementation starting with those of the highest promise while working to rapidly increase the potential of others. NETs cannot be seen as some magic technology by which we might achieve both net emissions reduction and $CO_2$ drawdown; they simply are not powerful enough for that. NETs really are the route for carbon drawdown, and they are to be operated alongside drastic reductions in ongoing emissions by whatever means possible and avoidance of new, additional emissions. In the absence of any one of these three elements, we will fail to achieve the Paris climate targets. It's that simple. Despite the triumphant announcements in the media and speculation in political circles about NETs, the reality today is that NETs may not be relied upon as some sort of superman technology that will save the world at the eleventh hour. Such things happen in comic books, not in reality.

But—and this is completely possible—what if I'm completely wrong about this role of NETs? After all, nobody knows whether some bright mind might discover something that is currently completely unknown. Let's consider that. First, this scenario assumes rescue by something we haven't yet got a clue about and that might or might not happen; do we really want to bet the future of our children and grandchildren on such an off-chance scenario? Or should we start reaping the benefits of responsible action on implementation of existing NETs already, which will at the same time bring improvements to biodiversity and ecosystem services; revitalize soils around the world; improve soil stability as well as water retention and availability in the landscape; provide greater food-crop security; reduce air pollution and toxic algal blooms; limit ocean acidification with fisheries benefits; provide benefits to public health, next-generation fuels, and so on. Surely, this is not a bad set of benefits to enjoy alongside the initial removal of carbon from the climate system, while we wait and see if /when an unknown superhero NET might emerge. Second, reality dictates that, if a new superhero NET emerges, it will be in the form of an invention or prototype of great promise, which then will need many years to a few decades of polishing and refining, testing and demonstrating, including environmental and public health risk assessments, before it might be considered ready for at-scale implementation. We cannot afford to twiddle our thumbs while we wait for such an eventuality to play out.

In sum, we need to cover emissions reductions and avoidance of new emissions growth by independent development and implementation streams (Chapter 3) and in addition focus all NETs activity we can reasonably deploy on addressing the excess 70–280 GtC removal requirement. If we can create a massive ramp-up of actually realized NETs capacity to a total average annual

**Table 4.4** Synthesis for the various NETs; basis for the numbers presented in Figures 4.12 and 4.13.

| | TRL | GtC/y Best | GtC/y Worst, realistic | Fresh water Best | Fresh water Worst | Land Best | Land Worst | Other impacts Best | Other impacts Worst | Other impact scores have been subjectively allocated on the basis of the following arguments: |
|---|---|---|---|---|---|---|---|---|---|---|
| RA | 9 | 3.30 | 0.14 | 370.00 | 1040.00 | 320.00 | 970.00 | 0.1 | 0.4 | Loss of biodiversity; possible destruction of wildlands and soil degradation |
| SCS | 9 | 3.90 | 0.11 | 0.05 | 1.00 | 0.05 | 1.00 | 0.1 | 0.2 | Possible impacts on balance of pristine ecosystems |
| BIOCH | 7 | 1.34 | 0.01 | 0.05 | 1.00 | 40.00 | 260.00 | 0.1 | 0.4 | Possible impacts on balance of pristine ecosystems; $CO_2$ release during pyrolysis; infrastructure for at-scale application |
| BECCS | 5 | 3.30 | 0.00 | 720.00 | 1040.00 | 10.00 | 380.00 | 0.2 | 0.6 | Loss of biodiversity; pressure on food crops and wild ecosystems; large industrial facilities with intricate infrastructure for CCS |
| DACCS | 8 | 1.35 | 0.14 | 10.00 | 300.00 | 0.05 | 1.00 | 0.1 | 0.4 | Large industrial facilities with intricate infrastructure for CCS |
| EW | 6 | 1.09 | 0.14 | 0.30 | 1.50 | 2.00 | 10.00 | 0.1 | 0.4 | Soil mineral unbalance; leaching out of toxins; transport of rock dust |
| PyCCS | 5 | 1.40 | 0.01 | 0.05 | 1.00 | 40.00 | 260.00 | 0.1 | 0.4 | Possible impacts on balance of pristine ecosystems; infrastructure for at-scale application |

*Continued*

**Table 4.4** *Continued*

| | TRL | GtC/y Best | GtC/y Worst, realistic | Fresh water Best | Fresh water Worst | Land Best | Land Worst | Other impacts Best | Other impacts Worst | Other impact scores have been subjectively allocated on the basis of the following arguments: |
|---|---|---|---|---|---|---|---|---|---|---|
| OA | 3 | 2.70 | 0.27 | 0.05 | 1.00 | 0.05 | 1.00 | 0.3 | 0.6 | Possible impacts on pristine ecosystems |
| OF | 3 | 12.00 | 0.14 | 0.05 | 1.00 | 0.05 | 1.00 | 0.2 | 0.9 | Possible upset of pristine ecosystems; subsurface oxygen depletion; uncertain longevity of C storage |
| NEH2 | 2 | 12.50 | 2.50 | 0.05 | 1.00 | 0.05 | 1.00 | 0.3 | 0.6 | As yet, very uncertain parameters, but possible impacts as OA |
| MP | 6 | 0.44 | 0.40 | 0.05 | 1.00 | 0.05 | 1.00 | 0.1 | 0.2 | Industrial infrastructure and transport requirements |
| AU | 2 | 1.00 | 0.00 | 0.05 | 1.00 | 0.05 | 1.00 | 0.1 | 0.6 | As yet, very uncertain parameters, but likely issues as OF, but on smaller scales |

| | $1-(1/TRL)$ | Values relative to max. of 12 | | Complement*, & rel. to 1100 | | Complement*, & rel. to 1000 | | Complement* | | |
|---|---|---|---|---|---|---|---|---|---|---|
| RA | 0.89 | 0.28 | 0.01 | 0.66 | 0.05 | 0.68 | 0.03 | 0.90 | 0.60 | 0.10 |
| SCS | 0.89 | 0.33 | 0.01 | 1.00 | 1.00 | 1.00 | 1.00 | 0.90 | 0.80 | 0.26 |
| BIOCH | 0.86 | 0.11 | 0.00 | 1.00 | 1.00 | 0.96 | 0.74 | 0.90 | 0.60 | 0.08 |
| BECCS | 0.80 | 0.28 | 0.00 | 0.35 | 0.05 | 0.99 | 0.62 | 0.80 | 0.40 | 0.06 |
| DACCS | 0.88 | 0.11 | 0.01 | 0.99 | 0.73 | 1.00 | 1.00 | 0.90 | 0.60 | 0.09 |

| | | | | | | | | | |
|---|---|---|---|---|---|---|---|---|---|
| EW | 0.83 | 0.09 | 0.01 | 1.00 | 1.00 | 1.00 | 0.99 | 0.90 | 0.07 | 0.01 |
| PyCCS | 0.80 | 0.12 | 0.00 | 1.00 | 1.00 | 0.96 | 0.74 | 0.90 | 0.08 | 0.00 |
| OA | 0.67 | 0.23 | 0.02 | 1.00 | 1.00 | 1.00 | 1.00 | 0.70 | 0.10 | 0.01 |
| OF | 0.67 | 1.00 | 0.01 | 1.00 | 1.00 | 1.00 | 1.00 | 0.80 | 0.53 | 0.00 |
| NEH2 | 0.50 | 1.04 | 0.21 | 1.00 | 1.00 | 1.00 | 1.00 | 0.70 | 0.36 | 0.04 |
| MP | 0.83 | 0.04 | 0.03 | 1.00 | 1.00 | 1.00 | 1.00 | 0.90 | 0.03 | 0.02 |
| AU | 0.50 | 0.08 | 0.00 | 1.00 | 1.00 | 1.00 | 1.00 | 0.90 | 0.04 | 0.00 |

*These values are presented as complements (1–value), to orient them for index calculation so that 0 = worst and 1 = best.

*Sources:* Top panel: Synthesis of the results reported by Shepherd et al. (2009), Yool et al. (2009), Oschlies et al. (2010), Gadikota et al. (2015), Smith et al. (2016a,b), Williamson (2016), Fuss et al. (2018), Henderson et al. (2018), Lawrence et al. (2018); Minx et al. (2018), Rau et al. (2018), Schmidt et al. (2018), Bastin et al. (2019), Levin (2019), IPCC (2019), National Academies of Sciences, Engineering, and Medicine (2019), Paustian et al. (2019), Bossio et al. (2020), Cook-Patton et al. (2020).

*Notes:* The scores given under "other impact" reflect my own subjective judgement, based predominantly on the qualitative arguments listed in the final column of the top panel. Index totals (as plotted in Figure 4.13) are calculated in the bottom panel using $T_{RL}$, which is $1-(1/TRL)$ so that the index gives greater weight to improvements between early TRL stages when innovation and creativity dominate advances, relative to improvement between later TRL stages when TRL largely depends on money invested; $C_P$, carbon removal potential normalized to a range between 0 and 1 by dividing all values in the top panel by the maximum value (12 GtC/y); $W_P$, which is the complement of freshwater requirement normalized to a range between 0 and 1 through division by 1100 km³/y (i.e., 1–Freshwater needs/1100); $L_S$, which is the complement of land surface area normalized to a range between 0 and 1 through division by 1000 million hectares (i.e., 1–Land needs/1000); and $I_{MP}$ which is the complement of the "other impacts" score (i.e., 1–Impact score). Complements were used in $I_{MP}$, $W_P$, and $L_S$ to ensure that the directions of all index components are harmonized to go from 0 = worst to 1 = best; in effect this inverts the reversed axes shown in Figure 4.12. Bottom panel: The index scores are calculated in a simple linear, equally weighted manner: $I = C_P \times T_{RL} \times I_{MP} \times W_F \times L_S$.

capture of 10 GtC, across our portfolio of options, then we'd be in with a shout of reaching the lower range of that removal requirement. We must pull out all the stops in optimizing some of the existing NETs further than anticipated in Figure 4.12, and in adding new independent NETs of high capacity. It'll be quite a ride, but we can still make it if we pull together. Just like the renewable energy sector and transport innovations, this NETs challenge reaches across many different disciplines and will lead to massive new industrial development, investment opportunities, technological advancement, and job creation, in addition to the broader benefits listed earlier.

# Key Sources and Further Reading

Antia, A.N., et al. Basin-wide particulate carbon flux in the Atlantic Ocean: regional export patterns and potential for atmospheric $CO_2$ sequestration. *Global Biogeochemical Cycles*, *15*, 845–862, 2001. https://agupubs.onlinelibrary.wiley.com/doi/abs/10.1029/2000GB001376

Appel, T., 2019. As in Chapter 3.

Bach, L.T., et al. $CO_2$ removal with enhanced weathering and ocean alkalinity enhancement: potential risks and co-benefits for marine pelagic ecosystems. *Frontiers in Climate*, *1*, 7, doi: 10.3389/fclim.2019.00007, 2019. https://www.frontiersin.org/articles/10.3389/fclim.2019.00007/full

Baker, D. Environmental groups respond to carbon capture funding pledge in UK Budget. *BDaily News*, March 13, 2020. https://bdaily.co.uk/articles/2020/03/13/environmental-groups-respond-to-carbon-capture-funding-pledge-in-uk-budget

Barrow, E. 300,000 hectares restored in Shinyanga, Tanzania—but what did it really take to achieve this restoration? *SAPIENS*, *7*, 2, p. 8, 2014. https://journals.openedition.org/sapiens/1542

Bastin, J.-F., et al. The global tree restoration potential. *Science*, *365*, 76–79, 2019. https://science.sciencemag.org/content/365/6448/76

Beal, C.M., et al. Integrating algae with bioenergy carbon capture and storage (ABECCS) increases sustainability. *Earth's Future*, *6*, 524–542, 2018. https://agupubs.onlinelibrary.wiley.com/doi/full/10.1002/2017EF000704

Beerling, D.J., et al. Farming with crops and rocks to address global climate, food and soil security. *Nature Plants*, *4*, 138–147, 2018. https://doi.org/10.1038/s41477-018-0108-y

Beerling, D.J., et al. Potential for large-scale $CO_2$ removal via enhanced rock weathering with croplands. *Nature, 583*, 242–248, 2020. https://www.nature.com/articles/s41586-020-2448-9

Bland, A. As oceans warm, the world's kelp forests begin to disappear. *YaleEnvironment360*, November 20, 2017. https://e360.yale.edu/features/as-oceans-warm-the-worlds-giant-kelp-forests-begin-to-disappear

Blaustein, R. Turning desert to fertile farmland on the Loess Plateau. *RE.THINK*, April 5, 2018. https://rethink.earth/turning-desert-to-fertile-farmland-on-the-loess-plateau/

Bonino, S. Carbon dioxide detection and indoor air quality control. *OHS Online*, April 1, 2016. https://ohsonline.com/articles/2016/04/01/carbon-dioxide-detection-and-indoor-air-quality-control.aspx

Bossio, D.A., et al. The role of soil carbon in natural climate solutions. *Nature Sustainability*, doi:10.1038/s41893-020-0491-z, 2020. https://www.nature.com/articles/s41893-020-0491-z

Boyd, P., and Vivian, C. Should we fertilize oceans or seed clouds? No one knows. *Nature, 570*, 155–157, 2019. https://www.nature.com/articles/d41586-019-01790-7

Brent, K., et al. International law poses problems for negative emissions research. *Nature Climate Change, 8*, 451–453, 2018. https://www.nature.com/articles/s41558-018-0181-2

Breyer, C., et al. Carbon dioxide direct air capture for effective climate change mitigation based on renewable electricity: a new type of energy system sector coupling. *Mitigation and Adaptation Strategies for Global Change, 23*, 43–65, 2019. https://doi.org/10.1007/s11027-019-9847-y

Buckingham, K. Chapter 16—beyond trees: restoration lessons from China's Loess Plateau. In: Song, L., et al. (eds.). *China's new sources of economic growth: reform, resources and climate change, Volume 1*, 379–395, ANU Press, The Australian National University, Canberra, Australia, 2016. http://press-files.anu.edu.au/downloads/press/n1906/pdf/ch16.pdf

Burns, W., and Rau, G.H. Can we tweak marine chemistry to help stave off climate change? *GreenBiz*, March 27, 2019. https://www.greenbiz.com/article/can-we-tweak-marine-chemistry-help-stave-climate-change

Butler, R.A. Earth has more trees now than 35 years ago. *Mongabay*, August 15, 2018. https://news.mongabay.com/2018/08/earth-has-more-trees-now-than-35-years-ago/

Carbon Engineering. *Direct air capture.* Accessed January 20, 2020. https://carbonengineering.com

Carrington, D.P. World's consumption of materials hits record 100bn tonnes a year. *The Guardian*, January 23, 2020. https://www.theguardian.com/environment/2020/jan/22/worlds-consumption-of-materials-hits-record-100bn-tonnes-a-year

Carroll, R. The wrong kind of trees: Ireland's afforestation meets resistance. *The Guardian*, July 7, 2019. https://www.theguardian.com/world/2019/jul/07/the-wrong-kind-of-trees-irelands-afforestation-meets-resistance?CMP=Share_AndroidApp_Gmail

Christophersen, T. Surprising benefits of an age-old land regeneration technique. *UN Environments Programme, Ecosystems Story*, March 27, 2019. https://www.unenvironment.org/news-and-stories/story/surprising-benefits-age-old-land-regeneration-technique

Christophersen, T. What would it really take to plant a trillion trees? *UN Environments Programme, Forests Blogpost*, August 23, 2019. https://www.unenvironment.org/news-and-stories/blogpost/what-would-it-really-take-plant-trillion-trees

Climeworks. *How it works.* Accessed January 14, 2020. https://climeworks.shop/how-it-works

Cohen, Y., and Rothman, D.H. Mechanisms for mechanical trapping of geologically sequestered carbon dioxide. *Proceedings of the Royal Society A, 471*, p. 10, 2015. http://doi.org/10.1098/rspa.2014.0853

Consoli, C. Bioenergy and carbon capture and storage. *Global CCS Institute Perspective*, 14, 2019. https://www.globalccsinstitute.com/wp-content/uploads/2019/03/BECCS-Perspective_FINAL_18-March.pdf

Cook-Patton, S.C., et al. Mapping carbon accumulation potential from global natural forest regrowth. *Nature, 585*, 545–550, 2020. https://www.nature.com/articles/s41586-020-2686-x

Daval, D. Carbon dioxide sequestration through silicate degradation and carbon mineralisation: promises and uncertainties. *Nature Partner Journals Materials Degradation, 2*, 11, p. 11, 2018. https://www.nature.com/articles/s41529-018-0035-4.pdf

De Villiers, C., et al. Carbon sequestered in the trees on a university campus: a case study. *Sustainability Accounting, Management and Policy Journal, 5*, 149–171, 2014.

DOE/Argonne National Laboratory. Iron fertilization, process of putting iron into ocean to help capture carbon, could backfire. *ScienceDaily*, June 12, 2013. https://www.sciencedaily.com/releases/2013/06/130612144833.htm

Donato, C., et al. Mangroves among the most carbon-rich forests in the tropics. *Nature Geoscience, 4,* 293–297, 2011. https://www.nature.com/articles/ngeo1123

EC (European Commission). HORIZON 2020—work programme 2014–2015, general annex G. Technology Readiness Levels (TRL). *Commission Decision C(2014),* 2014. https://ec.europa.eu/research/participants/data/ref/h2020/wp/2014_2015/annexes/h2020-wp1415-annex-g-trl_en.pdf

Evans, K. "We used to be leaders": the collapse of New Zealand's landmark ocean park. *The Guardian,* March 11, 2020. https://www.theguardian.com/environment/2020/mar/11/we-used-to-be-leaders-the-collapse-of-new-zealands-landmark-ocean-park

FERN. *Six problems with BECCS.* September 2018. https://www.fern.org/fileadmin/uploads/fern/Documents/Fern%20BECCS%20briefing_0.pdf

Field, J.L., et al. Robust paths to net greenhouse gas mitigation and negative emissions via advanced biofuels. *Proceedings of the National Academy of Sciences of the USA, 117,* 21968–21977, 2020. https://www.pnas.org/content/117/36/21968

Fitzsimmons, C. Save our minds by saving the world. *Sydney Morning Herald,* January 20, 2019. https://www.smh.com.au/environment/climate-change/we-can-save-our-minds-by-saving-the-world-20190117-p50rye.html

Frank, S., et al. Reducing greenhouse gas emissions in agriculture without compromising food security? *Environmental Research Letters, 12,* 105004, 2017. https://iopscience.iop.org/article/10.1088/1748-9326/aa8c83

Fromberg, A. Tasmanian kelp forests dying as water warms, dive operator Mick Baron says. *ABC News,* February 21, 2017. https://www.abc.net.au/news/2017-02-21/tasmanian-kelp-forests-dying-as-water-warms-dive-operator-says/8289300

Fuss, S., et al. Negative emissions—part 2: costs, potentials and side effects. *Environmental Research Letters, 13,* 063002, 2018. https://iopscience.iop.org/article/10.1088/1748-9326/aabf9f

Gadikota, G., et al. Carbonation of silicate minerals and industrial wastes and their potential use as sustainable construction materials. In: Jin, F., He, L.-N., and Hu, Y.H. (eds.). *Advances in $CO_2$ capture, sequestration, and conversion,* pp. 295–322, American Chemical Society, Washington, DC, 2015. doi: 10.1021/bk-2015-1194. ch012. https://pubs.acs.org/doi/pdf/10.1021/bk-2015-1194.ch012

Gambhir, A., and Tavoni, M. Direct air carbon capture and sequestration: how it works and how it could contribute to climate-change mitigation. *One Earth, 1,* 405–409, 2019. https://www.cell.com/one-earth/fulltext/S2590-3322(19)30216-7

Gattuso, J.-P., et al. Ocean solutions to address climate change and its effects on marine ecosystems. *Frontiers in Marine Science, 5,* 337, 2018. https://www.frontiersin.org/articles/10.3389/fmars.2018.00337/full

Geoengineering Technological Briefing. Direct air capture. *Geoengineering Monitor*, May 24, 2018a. http://www.geoengineeringmonitor.org/2018/05/direct-air-capture/

Geoengineering Technological Briefing. Artificial upwelling. *Geoengineering Monitor*, June, 2018b. http://www.geoengineeringmonitor.org/wp-content/uploads/2018/06/Geoengineering-factsheet-Upwelling.pdf

Geospace. *Storing carbon dioxide underground impacts microbial communities.* December 9, 2011. https://blogs.agu.org/geospace/2011/12/09/storing-carbon-dioxide/

Hambrett, M. How planting trees and grasses can help stabilise farmland in a changing climate. *ABC News*, August 3, 2018. https://www.abc.net.au/news/2018-08-03/how-trees-can-be-used-as-drought-busters/10069318

Haque, F., et al. Optimizing inorganic carbon sequestration and crop yield with wollastonite soil amendment in a microplot study. *Frontiers in Plant Science*, July 3, 2020. https://www.frontiersin.org/articles/10.3389/fpls.2020.01012/full

Hasegawa, T., et al. Food security under high bioenergy demand toward long-term climate goals. *Climatic Change*, 26 August, 2020. https://link.springer.com/article/10.1007/s10584-020-02838-8

Henderson, G., et al. Greenhouse gas removal. *Royal Society DES5563_1*, 2018. ISBN: 978-1-78252-349-9. https://royalsociety.org/-/media/policy/projects/greenhouse-gas-removal/royal-society-greenhouse-gas-removal-report-2018.pdf

Hickman, L. Timeline: how BECCS became climate change's "saviour" technology. *CarbonBrief, Features*, April 13, 2016. https://www.carbonbrief.org/beccs-the-story-of-climate-changes-saviour-technology

Ingall, E.D., et al. Role of biogenic silica in the removal of iron from the Antarctic seas. *Nature Communications*, 4, 1981, p. 6, 2013. http://dx.doi.org/10.1038/ncomms2981

Intergovernmental Panel on Climate Change working group 1, 2013. As in Chapter 2. IPCC, 2019. As in Chapter 3.

Jeffrey, S., et al. Biochar boosts tropical but not temperate crop yields. *Environmental Research Letters*, 12, 053001, 2017. https://iopscience.iop.org/article/10.1088/1748-9326/aa67bd

Köhl, M., et al. The impact of tree age on biomass growth and carbon accumulation capacity: a retrospective analysis using tree ring data of three tropical tree species grown in natural forests of Suriname. *PLOS ONE*, 12, e0181187, 2017. https://doi.org/10.1371/journal.pone.0181187

Lampitt, R.S., et al. Ocean fertilization: a potential means of geoengineering? *Philosophical Transactions of the Royal Society of London A: Mathematical, Physical and Engineering Sciences, 366,* 3919–3945, 2008. https://royalsocietypublishing. org/doi/pdf/10.1098/rsta.2008.0139

Laurance, B. The end of big trees? *The Conversation,* December 11, 2012. https:// theconversation.com/the-end-of-big-trees-11217

Lawrence, M.G., et al. Evaluating climate geoengineering proposals in the context of the Paris Agreement temperature goals. *Nature Communications, 9,* 3734, 2018. https://www.nature.com/articles/s41467-018-05938-3.pdf

Levin, K. How effective is land at removing carbon pollution? The IPCC weighs in. *World Resources Institute,* August 08, 2019. https://www.wri.org/blog/2019/08/ how-effective-land-removing-carbon-pollution-ipcc-weighs

Li, B., et al. Advances in $CO_2$ capture technology: a patent review. *Applied Energy, 102,* 1439–1447, 2013. https://medicine.hsc.wvu.edu/media/250441/applen_ 2013-2.pdf

Litynski, J. Office of Clean Coal and Carbon Management—charter for the group and overview of DOE program efforts. *Presentation of US DoE Office of Fossil Energy Workshop,* July 24, 2019. https://www.usea.org/sites/default/files/event-/Litynski_ DAC_FE_WORKSHOP2.pdf.

Liu, C.L.C., et al. Mixed-species versus monocultures in plantation forestry: development, benefits, ecosystem services and perspectives for the future. *Global Ecology and Conservation, 15,* e00419, 2018. https://www.sciencedirect.com/science/article/pii/S235198941830088X

Liu, J.D., and Hiller, B.T. Chapter 4.8—a continuing inquiry into ecosystem restoration: examples from China's Loess Plateau and locations worldwide and their emerging implications. In: *Land Restoration: Reclaiming Landscapes for a Sustainable Future,* pp. 361–379, Academic Press, 2016. https://www.sciencedirect. com/science/article/pii/B9780128012314000276

Lyra, A., de A. Sensitivity of the Amazon biome to high resolution climate change projections. *Acta Amazonica, 46,* 175–188, 2016. https://www.scielo.br/pdf/aa/ v46n2/1809-4392-aa-46-02-00175.pdf

Mac Dowell, N., et al. The role of $CO_2$ capture and utilization in mitigating climate change. *Nature Climate Change, 7,* 243–249, 2017. https://www.nature.com/articles/nclimate3231

Manning, D.A.C., et al. Carbonate precipitation in artificial soils produced from basaltic quarry fines and composts: an opportunity for passive carbon sequestration.

*International Journal of Greenhouse Gas Control, 17,* 309–317, 2013. https://www.sciencedirect.com/science/article/pii/S1750583613002156

Mathesius, S., et al. Long-term response of oceans to $CO_2$ removal from the atmosphere. *Nature Climate Change, 5,* 1107–1113, 2015. https://www.nature.com/articles/nclimate2729

Matter, J.M., et al. Rapid carbon mineralization for permanent disposal of anthropogenic carbon dioxide emissions. *Science, 352,* 1312–1314, 2016. https://www.researchgate.net/publication/303450549_Rapid_carbon_mineralization_for_permanent_disposal_of_anthropogenic_carbon_dioxide_emissions

MCi, *Mineral Carbonation International.* Accessed July 29, 2020. https://www.mineralcarbonation.com

Middleton, N., et al. *World atlas of desertification,* 182 pp., UNEP(017)/W927, 1997.

Ming, T., et al. Fighting global warming by climate engineering: Is the Earth radiation management and the solar radiation management any option for fighting climate change? *Renewable and Sustainable Energy Reviews, 31,* 792–834, 2014. https://www.sciencedirect.com/science/article/pii/S1364032113008460

Minx, J., et al. Negative emissions—part 1: research landscape and synthesis. *Environmental Research Letters, 13,* 063001, 2018. https://iopscience.iop.org/article/10.1088/1748-9326/aabf9b

Moosdorf, N., et al. Carbon dioxide efficiency of terrestrial enhanced weathering. *Environmental Science and Technology, 48,* 4809–4816, 2014. https://pubs.acs.org/doi/10.1021/es4052022

National Academies of Sciences, Engineering, and Medicine. *Negative emissions technologies and reliable sequestration: a research agenda.* The National Academies Press, Washington, DC, 2019. doi: https://doi.org/10.17226/25259. https://www.nap.edu/read/25259/chapter/1

Oldeman, L.R., et al. World map of the status of human-induced soil degradation: an explanatory note. *Global Assessment of Soil Degradation (GLASOD),* January 10, 1991. http://the-eis.com/elibrary/sites/default/files/downloads/literature/World%20map%20of%20the%20status%20of%20human-induced%20soil%20degradation_1991.pdf

Ore, A. Big old trees grow faster, making them vital carbon absorbers. *The Conversation,* January 17, 2014. https://theconversation.com/big-old-trees-grow-faster-making-them-vital-carbon-absorbers-22104?sfns=mo

Oschlies, A., et al. Climate engineering by artificial ocean upwelling: channelling the sorcerer's apprentice. *Geophysical Research Letters, 37,* L04701, 2010. https://agupubs.onlinelibrary.wiley.com/doi/full/10.1029/2009GL041961

Pan, Y., et al. A large and persistent carbon sink in the world's forests. *Science, 333,* 988–993, 2011. https://science.sciencemag.org/content/sci/333/6045/988.full.pdf

Pan, Y., et al. Research progress in artificial upwelling and its potential environmental effects. *Science China Earth Sciences, 59,* 236–248, 2016. https://www.researchgate.net/publication/286529279_Research_progress_in_artificial_upwelling_and_its_potential_environmental_effects

Parry, W. Could fertilizing the oceans reduce global warming? *LiveScience,* July 18, 2012. https://www.livescience.com/21684-geoengineering-iron-fertilization-climate.html

Paustian, K., et al. Soil C sequestration as a biological negative emission strategy. *Frontiers in Climate, 1,* 8, p. 11, 2019. https://www.frontiersin.org/articles/10.3389/fclim.2019.00008/full

Pearce, F. The natural solutions to climate change held in the ocean. *Chinadialogue, Ocean,* November 26, 2019. https://chinadialogueocean.net/11915-coastal-ecosystem-natural-solutions-climate-change/

Pendleton, L., et al. Estimating global "blue carbon" emissions from conversion and degradation of vegetated coastal ecosystems. *PLOS ONE, 7,* e43542, 2012. https://journals.plos.org/plosone/article/file?id=10.1371/journal.pone.0043542&type=printable

Permentier, K., et al. Carbon dioxide poisoning: a literature review of an often forgotten cause of intoxication in the emergency department. *International Journal of Emergency Medicine, 10,* 14, p. 4, 2017. https://www.ncbi.nlm.nih.gov/pmc/articles/PMC5380556/

Rau, G.H., et al. The global potential for converting renewable electricity to negative-$CO_2$-emissions hydrogen. *Nature Climate Change, 8,* 621–625, 2018. https://www.nature.com/articles/s41558-018-0203-0

Renforth, P., et al. The dissolution of olivine added to soil: implications for enhanced weathering. *Applied Geochemistry, 61,* 109–118, 2015. https://www.sciencedirect.com/science/article/pii/S0883292715001389

Renforth, P. and Henderson, G. Assessing ocean alkalinity for carbon sequestration. *Reviews of Geophysics, 55,* 636–674, 2017. https://agupubs.onlinelibrary.wiley.com/doi/abs/10.1002/2016RG000533

Rinaudo, T. FMNR frequently asked questions with Tony Rinaudo. *Food Security and Natural Resources team, World Vision Australia,* December 19, 2014. http://fmnrhub.com.au/frequently-asked-questions/#1

Rogelj, J., et al., 2018. As in Chapter 2.

Rohling, E.J., and Ortiz, J.D. We're killing our lakes and oceans. The consequences are real. *Undark*, February 6, 2018. https://undark.org/2018/02/06/dead-zones-oceans-lakes-coastal-seas/

Romanov, V., et al. Mineralization of carbon dioxide: a literature review. *ChemBioEng Reviews*, 2, 231–256, 2015. https://www.osti.gov/servlets/purl/1187926

Rost, B., et al. Carbon acquisition of bloom-forming marine phytoplankton. *Limnology and Oceanography*, 48, 55–67, 2003. https://aslopubs.onlinelibrary.wiley.com/doi/abs/10.4319/lo.2003.48.1.0055

Sanderman, J. et al. A global map of mangrove forest soil carbon at 30 m spatial resolution. *Environmental Research Letters*, 13, 12, 2018. https://iopscience.iop.org/article/10.1088/1748-9326/aabe1c/meta

Schaef, T. et al. Carbonate mineralization of volcanic province basalts. *International Journal of Greenhouse Gas Control*, 4, 249-261, 2010. https://doi.org/10.1016/j.ijggc.2009.10.009.

Schmidt, H.P. Biochar and PyCCS included as negative emission technology by the IPCC. *The Biochar Journal 2018*, Arbaz, Switzerland, 2018. ISSN 2297-1114. https://www.biochar-journal.org/en/ct/94

Schmidt, H.P., et al. Pyrogenic carbon capture and storage. *Global Change Biology, Bioenergy*, 11, 573–591, 2019. https://onlinelibrary.wiley.com/doi/10.1111/gcbb.12553

Scholz, S.M., et al. *Biochar systems for smallholders in developing countries: leveraging current knowledge and exploring future potential for climate-smart agriculture.* World Bank Studies, World Bank, Washington, DC, 231 pp., 2014. doi:10.1596/978-0-8213-9525-7. License: Creative Commons Attribution CC BY 3.0 IGO. https://openknowledge.worldbank.org/bitstream/handle/10986/18781/888880PUB0Box30lso0784070June122014.pdf?sequence=1

Schuiling, R.D. *The rate of olivine weathering, an expensive myth.* Smartstones: The Olivine Foundation. Accessed November 2019. http://smartstones.nl/the-rate-of-olivine-weathering-an-expensive-myth/

Schuiling, R.D., and Krijgsman, P. Enhanced weathering: an effective and cheap tool to sequester $CO_2$. *Climatic Change*, 74, 349–354, 2006. https://link.springer.com/article/10.1007/s10584-005-3485-y

Shackley S., et al., 2012. As in Chapter 3.

Shepherd, J.G., et al. Geoengineering the climate: science, governance and uncertainty. *Royal Society Policy document, 10/09.* The Royal Society, London, UK, September 2009. ISBN: 978-0-85403-773-5. https://royalsociety.org/~/media/Royal_Society_Content/policy/publications/2009/8693.pdf

Silva, L.C.R. Carbon sequestration beyond tree longevity. *Science, 355*, 1141, 2017. https://science.sciencemag.org/content/355/6330/1141

Slezak, M. Massive mangrove die-off on Gulf of Carpentaria worst in the world, says expert. *The Guardian*, July 11, 2016. https://www.theguardian.com/environment/2016/jul/11/massive-mangrove-die-off-on-gulf-of-carpentaria-worst-in-the-world-says-expert

Smith, P. Soil carbon sequestration and biochar as negative emission technologies. *Global Change Biology, 22*, 1315–1324, 2016a. https://onlinelibrary.wiley.com/doi/epdf/10.1111/gcb.13178

Smith, P, et al. Biophysical and economic limits to negative $CO_2$ emissions. *Nature Climate Change, 6*, 42–50, 2016b. https://www.nature.com/articles/nclimate2870

Snæbjörnsdóttir, S.Ó., et al. Carbon dioxide storage through mineral carbonation. *Nature Reviews Earth and Environment, 1*, 90–102, 2020. https://www.nature.com/articles/s43017-019-0011-8

Song, X.P., et al. Global land change from 1982 to 2016. *Nature, 560*, 639–643, 2018. https://www.nature.com/articles/s41586-018-0411-9.pdf

Stephenson, N.L., et al. Rate of tree carbon accumulation increases continuously with tree size. *Nature, 507*, 90–93, 2014. https://www.nature.com/articles/nature12914

Tanzer, S.E., and Ramirez, A., 2019. As in Chapter 3.

Taylor, A.P. Earth has room for 1 trillion more trees. *The Scientist, News and Opinion*, July 5, 2019. https://www.the-scientist.com/news-opinion/earth-has-room-for-1-trillion-more-trees-66103

The Climate Examiner. Researchers demonstrate the potential of carbon-negative hydrogen fuel. *Solutions*, July 11, 2018. http://theclimateexaminer.ca/2018/07/11/researchers-demonstrate-the-potential-of-carbon-negative-hydrogen-fuel/

Trias, R., et al. High reactivity of deep biota under anthropogenic $CO_2$ injection into basalt. *Nature Communications, 8*, 1063, 2017. https://www.nature.com/articles/s41467-017-01288-8

Turvey, C.C., et al. Hydrotalcites and hydrated Mg-carbonates as carbon sinks in serpentinite mineral wastes from the Woodreef chrysotile mine, New South Wales, Australia: controls on carbonate mineralogy and efficiency of $CO_2$ air capture in mine tailings. *International Journal of Greenhouse Gas Control, 79*, 38–60, 2018. https://www.sciencedirect.com/science/article/abs/pii/S1750583618302755

Veldman, J.W., et al. Where tree planting and forest expansion are bad for biodiversity and ecosystem services. *BioScience, 65*, 1011–1018, 2015. https://academic.oup.com/bioscience/article/65/10/1011/245863

Wahl, J.D. Human and planetary health: ecosystems restoration at the dawn of the century of regeneration. *Medium*, October 13, 2018. https://medium.com/@designforsustainability/human-and-planetary-health-part-iv-restoring-ecosystems-in-the-century-of-regeneration-d24bbfe37617

Wang, C., et al. Effects of biochar amendment on net greenhouse gas emissions and soil fertility in a double rice cropping system: a 4-year field experiment. *Agriculture, Ecosystems, and Environment, 262*, 83–96, 2018. https://www.sciencedirect.com/science/article/abs/pii/S016788091830166X

Weston, P., and Hong, R. Talensi farmer-managed natural regeneration project in Ghana—social return on investment report. *FMNR hub, World Vision Australia*, 2013. http://fmnrhub.com.au/wp-content/uploads/2013/09/SROI-Report_High-Resolution.pdf

White, R.P., et al. Pilot analysis of global ecosystems: grassland ecosystems. *World Resources Institute*, ISBN 1-56973-461-5, 2000. http://pdf.wri.org/page_grasslands.pdf

Wikipedia. *Ocean fertilization*. Accessed December 28, 2019. https://en.wikipedia.org/wiki/Ocean_fertilization

Williamson, P. Emissions reduction: scrutinize $CO_2$ removal methods. *Nature, 530*, 153–155, 2016. https://www.nature.com/news/emissions-reduction-scrutinize-co2-removal-methods-1.19318#/ref-link-10

Williamson, P. Guest post: 13 "ocean-based solutions" for tackling climate change. *CarbonBrief*, October 4, 2018. https://www.carbonbrief.org/guest-post-13-ocean-based-solutions-for-tackling-climate-change

Woolf, D., et al. Sustainable biochar to mitigate global climate change. *Nature Communications, 1*, 56, 2010. https://www.nature.com/articles/ncomms1053.pdf

Woolf, D., et al. Chapter 8—Biochar for climate change mitigation: navigating from science to evidence-based policy. In: Lal, R., and Stewart, B.A. (eds.) *Soil and Climate*, pp. 219–248, CRC Press, Taylor & Francis Group, Boca Raton, FL, 2018. ISBN 9781498783651. http://www.css.cornell.edu/faculty/lehmann/publ/Woolf%20et%20al%202018%20Biochar%20for%20Climate%20Change%20Mitigation.pdf

World Bank. *China—Loess Plateau watershed rehabilitation project*. World Bank, Washington, DC, 2003. http://documents.worldbank.org/curated/en/820471468769214283/China-Loess-Plateau-Watershed-Rehabilitation-Project

World Vision Australia, Food Security and Natural Resources Team. The spread of farmer-managed natural regeneration in Niger. *FMNR hub, World Vision Australia*, 2019. https://fmnrhub.com.au/projects/niger/#.XhPU662B2IY

Yamamoto, A., et al. Glacial $CO_2$ decrease and deep-water deoxygenation by iron fertilization from glaciogenic dust. *Climates of the Past, 15*, 981–996, 2019. https://cp.copernicus.org/articles/15/981/2019/cp-15-981-2019.html

Yool, A., et al. Low efficiency of nutrient translocation for enhancing oceanic uptake of carbon dioxide. *Journal of Geophysical Research, 114*, C08009, 2009. https://agupubs.onlinelibrary.wiley.com/doi/epdf/10.1029/2008JC004792.

Yu, J., et al. Sequestration of carbon in the deep Atlantic during the last glaciation. *Nature Geoscience, 9*, 319–324, 2016. https://www.nature.com/articles/ngeo2657

Yu, J., et al. More efficient North Atlantic carbon pump during the last glacial maximum. *Nature Communications, 10*, 2170, 2019. https://www.nature.com/articles/s41467-019-10028-z

Yu, J., et al. Last glacial atmospheric $CO_2$ decline due to widespread Pacific deep water expansion. *Nature Geoscience 13*, 628–633, 2020. https://www.nature.com/articles/s41561-020-0610-5

Zarandi, A.E., et al. Ambient mineral carbonation of different lithologies of mafic to ultramafic mining wastes/tailings—a comparative study. *International Journal of Greenhouse Gas Control, 63*, 392–400, 2017. https://www.sciencedirect.com/science/article/abs/pii/S1750583617303195

Zhang, D., et al. Reviews of power supply and environmental energy conversions for artificial upwelling. *Renewable and Sustainable Energy Reviews, 56*, 659–668, 2016. https://www.sciencedirect.com/science/article/abs/pii/S1364032115013064

# 5
# The Controversial One
## Solar Radiation Management

## 5.1. Setting the Scene

Solar radiation management, or SRM, seeks to manipulate the energy balance of climate by ensuring that less Incoming Short-Wave Radiation (ISWR) is absorbed. SRM is also known as solar geoengineering and spans a class of proposed measures that has been polarizing the community. Some see it as an essential means of keeping global warming within acceptable limits, while others see only grave drawbacks and dangers.

There are three key advantages. First, in contrast to emissions changes that control the greenhouse effect and associated climate feedbacks on initial response timescales of many years to decades (Chapters 3 and 4), SRM directly influences the amount of absorbed sunlight and affects temperature within months of implementation (National Research Council, 2015; MacMartin et al., 2018). Second, SRM-driven climate changes may be reversed on similarly short timescales. Third, certain SRM approaches are considerably cheaper to implement than most measures that result in net emissions changes (Moriyama et al., 2016).

Limiting the absorption of ISWR can be achieved by two means (Figure 5.1). First, measures may be taken in space, between Earth and the Sun, to reflect or disperse ISWR before it even hits Earth's atmosphere (section 5.2). Second, measures may be taken in Earth's atmosphere (section 5.3) or at the Earth surface (sections 5.4 and 5.5) to reflect ISWR. Some view SRM as a viable means of early and/or temporary intervention to limit warming while greenhouse gas concentrations may be brought under control.

But there are major differences of opinion about the wisdom behind this approach. For example, SRM by itself does not limit greenhouse gas concentrations, and thus does not help at all in limiting ocean acidification. And there is a risk that temperature management by SRM would take the

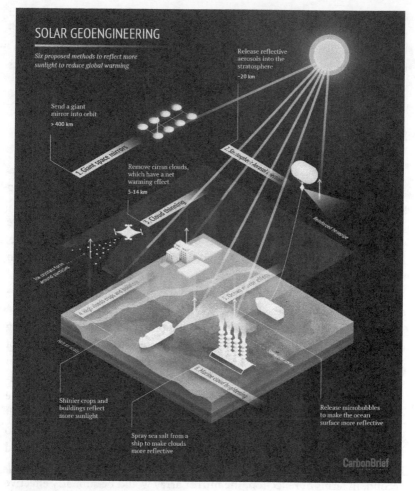

**Figure 5.1**   SRM options.

*Source*: By Rosamund Pearce for Carbon Brief. Used by permission. https://www.
carbonbrief.org/explainer-six-ideas-to-limit-global-warming-with-solar-geoengineering

pressure off the need to implement measures to reduce greenhouse gas concentrations. Moreover, termination of temporary SRM application or an un-
intended ending to permanent SRM application—for example because of
economic reasons or because of war—could result in a shock re-adjustment
of the climate. These, and other downsides and challenges to different SRM

approaches, are discussed in section 5.7. In section 5.7, I also present a simple index that scores SRM approaches relative to one another according to their radiative reduction potential, environmental impacts, and technological readiness level (TRL), along with a table that presents the numbers used to develop that index.

## 5.2. Space-based SRM

The main space-based SRM technology that has been proposed involves sending a giant mirror—or fleet of mirrors—into orbit to reflect incoming sunlight before it reaches Earth (Figure 5.1).

Given a global ISWR average of 340 W/m$^2$ at the top of the atmosphere and a natural reflection of 100 W/m$^2$ of this by clouds and at the surface, Earth currently gains some 240 W/m$^2$ of ISWR. The level of net increase in the radiative forcing of climate since the start of the industrial revolution is about 2.25 W/m$^2$, mainly because of retention of outgoing long-wave radiation by greenhouse gases.

One would need a mirror blocking 1.8% of the solar radiation coming to Earth to offset the radiative forcing of climate caused by a doubling of $CO_2$ levels, which equates to 3.7 W/m$^2$ (Lenton and Vaughan, 2009). In consequence, we'd need a mirror blocking roughly 1.1% of the solar radiation coming to Earth to offset the changes in the energy budget of climate change that we have caused over the last 250 years or so. In those terms, it almost sounds feasible. But let's just put this into a bit of context. First, to block 1.1% of the incoming solar radiation, a mirror array would be needed that is almost 3 million km$^2$ in size (scaled after Lenton and Vaughan, 2009). Second, in terms of reflection, the mirror array would be similarly effective to the enormous North American and Eurasian ice sheets of the last ice age together—these combined ice sheets were roughly 2 times larger than the modern Antarctic ice sheet. These scale arguments portray the space mirror idea for what it is: a task of truly gargantuan proportions.

And then we have to look at projections of climate forcing increase by 2100. By then, we will be adding up to 0.75 W/m$^2$ for the lowest emissions scenario and some 6.25 W/m$^2$ for the most pessimistic scenario. Given that the current emissions pathway sits at or even just above the most pessimistic scenario, we will most likely need a mirror that reflects more than 4% of all incoming sunlight by 2100 to offset the anticipated 8.5 W/m² of human-caused climate forcing by 2100, relative to pre-industrial times. If 1.1% sounded like a gargantuan task, then 4% surely starts to sound impossible, especially because we don't really know yet how it should be done.

Installing a space mirror, or fleet of mirrors, in orbit poses a huge technological challenge. It will require the most advanced technology of all climate solutions, and the costs will far surpass those of all other SRM methods. On a positive note, the mirrors might be seen as the least environmentally disruptive method because they are in space. But that, of course, requires that we think only about their physical presence. It ignores impacts of reduced insolation on the biosphere—most notably on photosynthesis—and on the hydrological cycle, governing the potential intensities of droughts and floods. These and other considerations will be discussed in more detail in section 5.5.

## 5.3. Atmosphere-based SRM

There are several concepts for atmosphere-based SRM. The main ones are stratospheric aerosol injection, marine cloud brightening, cirrus cloud reduction, ocean sulfur cycle enhancement, and iron salt aerosol injection. Some of these suggested approaches are at a higher level of developmental maturity than others. Stratospheric aerosol injection is perhaps the furthest along because it does not require a huge technological advance; essentially, we have all the technical know-how to make it a reality today. But even stratospheric aerosol injection is not fully understood in terms of either its environmental impacts or its large-scale potential. All SRM methods, including stratospheric aerosol injection, remain to be fundamentally tested in field trials before large-scale trials or implementation can even be considered.

## 5.3.1. Stratospheric Aerosol Injection

The concept of stratospheric aerosol injection, releasing chemicals at altitudes of typically about 20 kilometers, was inspired by major volcanic eruptions that managed to eject a plume of ash and sulfur dioxide ($SO_2$) well into the stratosphere. The eruption of Mount Pinatubo in 1991 (Figure 5.2),

**Figure 5.2**  The eruption of Mount Pinatubo, June 12,1991, 08:51h.
*Source*: Photograph by Dave Harlow, United States Geological Survey. http://vulcan.
wr.usgs.gov/Volcanoes/ Philippines/Pinatubo/images.html

which came just when the scientific community had finished its preparations
for extensive measurements of the next major volcanic eruption, has become
the posterchild test-case for stratospheric aerosol injection (e.g., Hoppe,
1992; Self et al., 1996). It injected some 20 million tons of $SO_2$ aerosol into the
stratosphere, which combined with water to form sulfuric acid aerosols (that
is, hydrogen-sulfate, or $H_2SO_4$, which dissociates in water to provide sulfate
ions; $SO_4^{2-}$). Such aerosols in the stratosphere are very bright and highly reflective to sunlight.

The resultant global haze of aerosol caused a 2–4% decline in total solar
radiation to Earth's surface. It also affected how directly the sunlight reached
the surface; because of light scattering and reflection within the haze, there
was a 20–30% reduction in the amount of sunlight that reached Earth's surface directly—instead, much of the light was diffuse. From 1991 through
1993, global average temperatures dropped by about 0.5°C. But at the
same time, a detrimental consequence of stratospheric $SO_2$ injection came
to light: the stratospheric ozone layer was attacked. In consequence, mid-
latitude ozone concentrations reached a record low in 1992–1993, and the

so-called ozone hole in the southern hemisphere increased to a record size in 1992.

Stratospheric aerosol injection by means of planes or high-altitude balloons aims to mimic the cooling effects of large volcanic eruptions. For particularly relevant presentations of methods and impacts, see Robock et al. (2009), Jones et al. (2011, 2017), Curry et al. (2014), and MacMartin et al. (2018). Some studies suggest that it might even be possible to restore rainfall patterns to their pre-industrial averages, though other studies suggest that releasing aerosols only in the northern hemisphere may dangerously reduce rainfall in places like the African Sahel. Other impacts depend on the regions of stratospheric aerosol injection and include reductions in the frequency and intensity of storms over individual ocean basins. Alternatively, intense injection in just one hemisphere may result in the opposite, namely an increase in the number of storms. There are also suggestions that climate extremes, such as heatwaves or extreme precipitation, might be alleviated by injection over specific land regions.

While important potential applications are apparent, it is also clear that such applications need to be assessed thoroughly, in a holistic manner, to understand what would happen in other regions at the same time. After all, it would be undesirable for one nation to undertake stratospheric aerosol injection to improve conditions over their territory, if that unwittingly has severely detrimental impacts somewhere else. Or, heaven help us, for a nation to do it with the explicit aim of negatively affecting conditions over another region, for example to destroy crops. This brings us squarely to potential weaponization—enter the specter of climate-manipulation wars. We'll discuss that some more in section 5.5.

The substance to be released at altitude remains a topic of discussion. $SO_2$ gas combines with water to produce sulfuric acid, but this process may be too inefficient because slow conversion to sulfuric acid tends to produce aerosol particles that are too large. Direct injection of sulfuric acid vapor has been proposed instead, as this would be more efficient than $SO_2$ injection (Pierce et al., 2010). Hydrogen sulfide ($H_2S$) might be an alternative substance for stratospheric aerosol injection (Robock et al., 2009). You may know this gas from its distinct stink of rotten eggs, when inhaled in small doses. In somewhat higher doses, it is lethal. $H_2S$ would oxidize rapidly to form $SO_2$, which would then react with water to form sulfuric acid droplets. The key advantage of $H_2S$ is that it has a lower molecular weight than $SO_2$, so that roughly only half the weight would need to be lifted to produce the same amount of sulfate aerosol. Because $H_2S$ is so dangerous, however, it is probably not the best choice. Yet, note that $SO_2$ is also classed as a hazardous substance.

There are significant concerns that injecting sulfate aerosol into the stratosphere to alleviate global warming would at the same time introduce new climate risks, including ozone loss and heating of the lower tropical stratosphere. The latter would increase water vapor concentrations and thus cause further ozone loss as well as surface warming (Keith et al., 2016). As an alternative, therefore, some suggest injection of calcite ($CaCO_3$) aerosol particles. Calcite aerosols also have the capacity to increase reflection of incoming sunlight, but have the added benefit of helping to increase ozone levels back toward pre-industrial values. Further alternatives are soot aerosols injected at very high altitudes of 44–50 kilometers, or aluminum aerosols deployed at about 50 kilometer altitudes (Lenton and Vaughan, 2009). However, there may be undesirable side effects with these in terms of changes in stratospheric water vapor contents and stratospheric dynamics (Lawrence et al., 2018).

Lenton and Vaughan (2009) find that sulfate aerosol injection could theoretically be done to such an extent that about 3.7 W/m$^2$ reduction might be achieved in ISWR reaching the surface, while the impacts of soot and aluminum aerosols are more likely to range at around 1.9 W/m$^2$ and 0.5–1 W/m$^2$, respectively. A wider range for sulfate aerosols of 2–5 W/m$^2$ is given by Lawrence et al. (2018).

While many stratospheric aerosol injection proposals concern cooling of the entire planet, some studies have suggested targeting only the Arctic, with the objective of preventing the disappearance of summer sea ice and preserving the Greenland ice sheet. Robock et al. (2009) state:

> A disadvantage of Arctic injection is that the aerosols would only last a few months rather than a couple years for tropical injection … An advantage is that they would only need to be injected in spring [for] strongest effects over the summer. They would have no effect in the dark winter. One important difference between tropical and Arctic injections is that the [base of the stratosphere] is at about 16 km in the tropics but only about 8 km in the Arctic.

If stratospheric aerosol injection over the Arctic could indeed halt or even reverse sea-ice reduction, or avoid Greenland ice reduction, then a feedback process will come into play; that is, the high albedo (reflectivity) of these ice masses themselves will further help to limit Arctic warming.

As yet, all we think we know about stratospheric aerosol injection comes from models. Field trials are desperately needed to validate the processes and impacts suggested by the modeling studies and also to figure out the true costs of this method. For example, by critically assessing costs, Moriyama et al. (2017) found that previous studies had underestimated the cost of aircraft-based sulfate aerosol injection and that the annual cost to achieve radiative reduction of 2 W/m² could reach $10 billion, based on newly designed specialist aircraft. Costs would be even higher if existing aircraft were used. They also determined that a sulfate aerosol injection operation would likely require a fleet of approximately 1000 aircraft because of the high altitude required for injection. In short, field trials are essential to get a better grip on both the potential cooling and environmental impacts as well as the financial costs.

## 5.3.2. Marine Cloud Brightening

A second atmosphere-based SRM approach is marine cloud brightening or marine cloud seeding. It is also inspired by natural processes. Over oceans and seas, salty sea spray is an important cause of cloud nucleation; that is, the salt particles in the spray facilitate condensation of atmospheric water vapor. More droplets mean bigger, whiter, brighter clouds. These bright clouds reflect incoming sunlight, especially the so-called stratocumulus clouds. The marine cloud brightening approach aims to artificially spray fine mists of sea water into such clouds, accelerating the nucleation process and making the clouds bigger and brighter. This increases the amount of reflected sunlight, which in turn has an intensified cooling effect. A key advantage of this method is that it utilizes natural sea water, rather than potentially hazardous chemicals. A limitation is that it will work only in regions where relative humidity is high already.

Technological innovation is needed for sufficiently reliable generation of optimally sized particles to sustain continuous injection into low-lying marine clouds, given that the particles will precipitate naturally. Getting the particle size, as well as the injection amount, just right is of critical importance because under the wrong conditions the radiative forcing effect could go to zero, or even to positive values that would result in warming (Alterskjær and Kristjánsson, 2013).

Marine cloud brightening targets low-level clouds close to the sea surface. It therefore has more region-specific impacts than high-level aerosol injection into the very vigorously circulating stratosphere, where aerosols mix on hemispheric scales within months. Marine cloud brightening might, in theory, be used to impose targeted cooling just over coral reefs, to only cool one particular ocean region to try and reduce tropical cyclone strength in that area, to reduce polar temperatures and ice cover, and possibly to affect crop yields in downwind areas. Mostly these applications remain just ideas because, so far, there has been insufficient in-depth research (Dunne, 2018). Still, modeling experiments have borne out intriguing potential for such regional applications (Latham et al., 2014). The main impediments to such assessments lie in insufficient understanding of aerosol-cloud interactions and of the upscaling from cloud microphysics to regional cloud cover and climate impacts. In other words, the much-needed detailed modeling of marine cloud brightening so far remains impeded by a need to better understand and represent processes on a fundamental level. And beyond the modeling, field trials will be needed to see how theory translates into real-world results. For these, criteria have been drawn up already on how to determine success, manage international oversight and/or governance, and assess outcomes, if initial tests are deemed successful (Wood and Ackerman, 2013).

There is some field evidence that indicates the feasibility of marine cloud brightening, based on ship emissions that caused intensification of cloud formation (Figure 5.3). This unintentional experiment by the shipping industry is thought to have suppressed global warming by as much as 0.25°C, relative to what would have happened without it (Temple, 2018). In section 2.2, we encountered this cooling effect on a global combustion scale, and we also saw that cleaning up this air pollution would result in further warming. The emissions-aerosol effect of shipping has so far offset the warming caused by its greenhouse gas emissions. But this will change now that ship emissions of $SO_2$ and $NO_x$ are being targeted for reduction by international regulations to reduce air pollution, because of its harmful impacts on health and on water and soil acidification (Fuglestvedt et al., 2009).

Different modeling studies using different scenarios suggest that reductions in the radiative balance of climate between 0.8 and 5.4 $W/m^2$ are possible by enhancing the reflection of incoming sunlight through marine

**Figure 5.3** Cloud formation caused by ships' emissions.
*Source*: NASA/Goddard Space Flight Center Scientific Visualization Studio.

cloud brightening (Latham et al., 2008; Jones et al., 2009; Rasch et al., 2009; Lawrence et al., 2018). About 3.7 W/m$^2$ of negative radiative change would quantitatively offset the radiative increase caused by a doubling of atmospheric $CO_2$ levels. Engineering calculations suggest that a fleet of 1500 specialized, wind-driven, 300-ton vessels might suffice to achieve this much reduction in incoming solar radiation, though much more research is needed into the "design of an efficient spray generator, ... droplet life and dispersion, ... the present distribution of cloud condensation nuclei, a limited-area field experiment, and rigorous meteorological modeling ... [including assessment of] any adverse effects" (Salter et al., 2008). However, several caveats apply to the assumptions made in the radiative-change calculations underpinning those estimates, so that 1.7 W/m$^2$ reduction may be a more realistic outcome for that engineering effort (Lenton and Vaughan, 2009).

Conceptually, the climatic impacts of marine cloud brightening should switch on or off within 5 or 10 years of starting or ending the application because the particles that are used precipitate naturally on relatively short timescales. More complete modeling, however, has raised the possibility of serious complications to this idea (Jones et al., 2009). Especially for cloud modification in the South Atlantic, it was found that changes in the land-sea temperature contrasts may reduce rainfall and tree growth in the Amazon rain forest, reducing the forest's $CO_2$ uptake. In consequence, atmospheric $CO_2$ levels would remain higher, causing a stronger greenhouse effect than would exist if the South Atlantic cloud modifications were not applied. This effect would extend long beyond a switch-off of cloud modification. This substantially complicates any proposals for using marine cloud brightening to temporarily cool the planet while negative emissions methods are brought into action. To confuse matters further, no dramatic impacts on rain forests were found in other modeling studies, which used different cloud modification strategies (Rasch et al., 2009). Regardless, one thing is clear: the stakes are too high; even the smallest risk of major impacts on rain forests cannot be ignored. This issue needs resolving before large-scale implementation can be considered.

Model-based comparison between stratospheric aerosol injection and marine cloud brightening under a so-called business-as-usual emissions scenario has revealed some further issues (Jones et al., 2011). Both SRM approaches were applied in the model to such an extent that they offset the emissions-caused global average temperature change relative to the period 1990–1999. Both then show that in the case of no global average temperature change, there still is an important precipitation change; in fact, global average precipitation was found to be reduced by about 2% for both SRM approaches. In other words, while emissions rise drives an increase in global average precipitation, both SRM approaches more than offset this increase, leaving an overall 2% reduction in global average rainfall. Thus, SRM in the model managed to almost perfectly offset the global average temperature change, but caused notable rainfall reduction, relative to the control period of 1990–1999. This happens because climate responds differently to a change in incoming long-wave radiation than to a change in outgoing short-wave radiation.

In terms of spatial patterns, the comparison study revealed that stratospheric aerosol injection produces cooling that is distributed in a roughly similar pattern to the warming associated with emissions increase. A similar trend of opposing influences was found for precipitation changes, even though the exact magnitudes differed by enough to give rise to the global average reduction mentioned in the previous paragraph.

In the case of marine cloud brightening, however, the spatial patterns of change in temperature and precipitation were not similar and opposite. In some regions, marine cloud brightening even strengthened the temperature and precipitation anomalies caused by increasing emissions. In terms of choosing between the two methods, stratospheric aerosol injection might therefore be seen as less risky than marine cloud brightening. But let's not jump rashly to any conclusions on the basis of one study and instead do the modeling and field trials needed to build a proper foundation of understanding from which to develop a strategy toward potential implementation.

Rasch et al. (2009) make a further important point. Their model suggests that global averages of temperature, precipitation, and sea ice might be restored to present-day values using marine cloud brightening, but not simultaneously, and also with changed regional patterns relative to those of today. One might argue that this is undesirable. But the study then puts these differences from present-day values in a critical context: the regional patterns appear to change even more when allowing ongoing emissions in the absence of marine cloud brightening. Thus, we may face a choice between two evils. Either we apply marine cloud brightening and create some undesirable environmental changes, or we do not apply it and end up with even larger undesirable environmental changes. No doubt, similar arguments will apply to stratospheric aerosol injection. Careful establishing of the pros and cons of either scenario will help with deciding which is least harmful and how the scenarios compare with the impacts of doing nothing. The process of making decisions thus requires thorough and transparent answers to some pretty profound ethical and political questions.

While seawater spray is the main approach investigated for marine cloud brightening, there are some other methods, although most depend heavily on technological breakthroughs. One, however, is less technologically challenging. It centers around ocean iron fertilization, similar to the NET we discussed in Chapter 4. Iron fertilization in targeted regions can increase algal productivity, which increases the natural marine sulfur cycle, notably the algal release of dimethyl sulfide, or DMS. Most of us know DMS and its oxidation products from the characteristic smell of sea air; what we smell is not the sea itself, but a byproduct of algae. DMS oxidizes into a range of compounds, one of which is sulfuric acid, which promotes cloud droplet formation. Thus, iron fertilization offers potential to not only act as a NET, but also as a simultaneous SRM. Using iron fertilization in the Southern Ocean might therefore be considered to drive regional cooling and help slow down the loss of ice around Antarctic, even if its NET potential doesn't work out.

Finally, one further, nascent SRM proposal must be mentioned that involves manipulation of specific clouds. In this case, the focus lies on thinning of cirrus clouds. Cirrus clouds are thin, feathery clouds at high altitudes. They consist of ice crystals and they have two influences: they reflect some sunlight, and they absorb large amounts of outgoing long-wave radiation. Hence, this SRM is only partly a regulator of ISWR; the other, dominant part concerns modification of outgoing long-wave radiation. The net effect is that they warm the planet. There are proposals for removing them by injection of solid aerosol particles (for example, desert dust), which would result in cooling. A modeling study has suggested that cirrus cloud thinning might lead to a reduction in the radiative balance of climate by 1.6 $W/m^2$, producing cooling (Muri et al., 2014). But too much seeding might result in the opposite because it would thicken the cirrus clouds. This SRM method evidently is too poorly understood yet to even start considering the negative environmental impacts, costs, feasibility, and inevitable ethical and political issues that would surround its implementation.

## 5.3.3.  Iron Salt Aerosols

A third major SRM approach to be mentioned in this section is quite a bit broader than just an atmosphere-based SRM. Instead, it integrates marine cloud brightening with a wide range of additional benefits. This approach involves Iron Salt Aerosols, or ISA (Oeste et al., 2017). The method was inspired by the impacts of natural wind-blown dust in the atmosphere. Iron chloride is naturally present in the atmosphere as a salt aerosol, which results from reactions between hydrogen chloride and wind-blown iron oxides. The aerosol acts as cloud condensation nuclei, and thus drives marine cloud brightening. Similar to the marine cloud brightening approach discussed in section 5.3.2, the ISA approach aims to increase these aerosol levels in the atmosphere to help cloud brightening and cool the planet.

The ISA method proposes direct release of iron chloride, by balloon, ship, or plane, or additive-based production of iron oxy-hydroxides in industrial combustion processes; notably, fossil-fuel power stations. The iron oxy-hydroxides would quickly react to form iron chloride upon release of the exhaust gases. A multitude of beneficial impacts then develops in theory, including cloud nucleation and thus brightening; iron-driven enhancement of algal productivity that then causes net carbon drawdown as well as increased DMS release and further marine cloud brightening; chlorine-driven conversion of atmospheric methane into $CO_2$ with a lower global warming potential; chlorine-driven conversion of low-level ozone into oxygen, which reduces global warming influences; and chlorine-driven destruction of hydrochlorofluorocarbon gases that have high global warming potentials. The reported list of proposed benefits actually is even longer than this (ISA Australia, 2019).

The main problem with the ISA method is that it has been only qualitatively or semi-quantitatively described on the basis of relationships and reactions, without a rigorous quantitative understanding of the true impacts with fluxes and longevity assessments, combined impacts and potential cancellations between different influences, and possible adverse environmental impacts. Comprehensive assessment, including field trials, is needed to assess the real potential for radiative change of the ISA method, before large-scale implementation can be considered.

Regardless, the fundamental thought behind this method appears promising, and broad-spectrum approaches such as ISA may be just what the doctor ordered for dealing with the climate-change problem from several directions at the same time; if it works. As such, ISA presents a good candidate for high-priority research. In this research, attention is needed to magnetite (and iron oxide) releases during fossil-fuel and biomass burning, which would accompany the industrial combustion processes during which the ISA method foresees additive-based production of iron oxy-hydroxides. Such releases are important because magnetite aerosols have a warming effect on climate; fossil-fuel-based routes for ISA application may therefore have negative byproducts that could partly offset the ISA benefits. As stated by Matsui et al. (2018), "combustion iron is a ... more complex climate forcer than previously thought, and ... plays a key role in the Earth system." Clearly, all processes involved in this complex behavior need to be accounted for in assessments of the ISA hypothesis, to determine its true net potential.

## 5.4. Sea-Surface SRM

A few SRM approaches have been proposed that focus on increasing the albedo, or reflectivity, of the sea surface. In this section, I will briefly discuss two main concepts that have gained some traction. Yet, both remain at a stage where much further research is needed to reach a proper understanding of their true potential, environmental impacts, feasibility, and real-life challenges and performance in field-trials, before larger-scale implementation can be considered.

One approach involves the use of a fleet of vessels to create a sea foam of millions of tiny microbubbles on the ocean surface and/or adding chemical foaming agents to the wakes of ships (Evans et al., 2010). Incidentally, this approach is not limited to sea surfaces only, but may be applied in any standing water body (Seitz, 2011). The idea is to make dark, absorbent surface water more reflective to incoming sunlight. The effect would last only as long as the bubbles survive and/or are replaced by new ones. Initial concerns that increased reflection of sunlight might reduce the amount of light available for photosynthesis, and thus reduce algal productivity, are no longer considered alarming. Instead, key concerns now center on the environmental impacts of any chemicals used and on the amount of energy needed to create and sustain the microbubbles over a large enough surface area to significantly modify the net amount of ISWR.

A model-based comparison (Crook et al., 2016) found that ship wakes from existing international shipping (Figure 5.4) can indeed be modified to give a noticeable cooling effect of about 0.5°C, provided that the wake area could be tripled. In terms of the energy balance of climate, the impact was between 0.3 and 1.5 W/m$^2$ extra reflection of ISWR. This tripling of the wake area can be achieved by making the wakes survive longer, which requires the addition of chemicals to ensure that near-surface bubble lifetimes are extended to about 10 days. These chemicals, called surfactants, would obviously need to be environmentally harmless. But there are other factors to be considered too. Microbubbles at the sea surface enhance air-sea gas exchange, including the uptake of $CO_2$ and oxygen into and the release of DMS out of the sea water. In contrast, surfactants inhibit gas exchange. Greatly increasing surfactant additions may therefore considerably affect the environment in an indirect, adverse manner. Whether or not the microbubble effect can offset the surfactant effect remains a critical open question.

**Figure 5.4**  Wake of an ocean liner.
*Source*: Cruise Bruise Blog (2017). https://cruise-bruise.com/blog/2017/05/25/
reality-check-suicide-by-jumping-overboard-just-theatrical-romance-fiction/

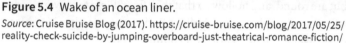

Just as marine cloud brightening and ISA applications might be regionally focused on polar regions to help reduce the loss of ice, the ship wake
study showed promise in protecting Arctic sea ice. In the extended ship
wake study, this Arctic focus arose automatically from the fact that there is
much more shipping traffic in the northern hemisphere than in the southern
hemisphere.

Model-based comparison between temperature and precipitation patterns expected under ongoing climate change and those expected in the extended ship wake experiment revealed that the scale of impacts of ongoing
climate change exceeds the scale of impacts of ship wake extension (Crook
et al., 2016). In detail, different patterns were found to affect different regions. Thus, we are back in the same ethical discussion that we encountered
with marine cloud brightening: which is the lesser of two evils? And who
makes that decision? Can one nation decide to proceed, even if concentrated
detrimental impacts are expected in another nation?

\* \* \*

Further sea-surface, and more specifically sea-ice-surface, SRM proposals
involve measures to increase the albedo of sea-ice, or to artificially thicken
sea ice, so that more of it survives through the summer, reflecting more

sunlight by itself. In other words, the idea behind these measures is to trigger a positive feedback process by cooling the polar regions, which promotes the growth of sea ice, which causes further cooling, which promotes further sea-ice recovery, and so on, until a new energy balance is reached with colder polar climates.

Albedo increase at polar latitudes might be achieved through the dispersal of "reflective materials such as [very fine and hollow] glass microspheres on young, low-reflectivity sea ice ... This way the ice may be conserved and converted over time into highly reflective multiyear sea ice. [This] can cool the Arctic significantly, and can rebuild Arctic ice area and volume, hence reducing Arctic as well as global temperature rise" (Field et al., 2018). The hollow glass microspheres considered in this study are chemically and physically like sand, but are round, and hollow so that they float. Field tests were undertaken on lakes using bright, white, hollow, and non-toxic glass K1 microspheres with an average diameter of 0.065 millimeters.

Only thin coverings of single or just a few layers would be needed to achieve the desired effect, and the field trials found no detrimental environmental impact of the spherules after the experiments. But small field trials are not enough when it comes to establishing the risks of applying artificial products at very large scales to restore Arctic ice. Such follow-up work will need to cover the ultimate fate of the products in polar environments, as well as the potential impacts on key species, such as marine mammals.

There is also a proposal to thicken sea-ice cover directly, by spraying or pumping water on top of young ice. Sea ice normally thickens by freezing of sea water onto its base. Because ice acts like an insulator, however, this process is slower than freezing of water on top of the ice, where it is in direct contact with the atmosphere, Thus, spraying sea water onto young sea ice would thicken the sea-ice layer more quickly than natural basal growth would. One proposal is to do exactly this, by building 10 million wind-powered, buoy-mounted pumps to cover 10% of the Arctic ice cap, at an estimated cost of $500 billion (Desch et al., 2017). The authors estimate that application of their sea-ice thickening approach to such an extent might increase the yearly average thickness of sea ice across the Arctic by about 1 meter every 10 years and increase ice volume sufficiently to counteract the current trend of Arctic sea-ice loss. The study mentions that covering the entire Arctic with ice would account for a globally averaged extra reflection of about 0.14 $W/m^2$.

As yet, however, the sea-ice thickening method is in the early stages of development. Questions remain, for example, about the environmental impacts of producing and then deploying very large numbers of buoys across the Arctic. And pilot studies are needed to iron out complexities such as equipment operability in Arctic conditions and to evaluate the method's real-life potential. Interestingly, such pilot studies could have additional benefits by restoring and strengthening sea-ice habitats for polar bears and marine mammals.

## 5.5. Land-Surface SRM

Land-surface SRM approaches have been summarized in detail under the collective name of regional land radiative management, abbreviated as $LRM_{reg}$ (Seneviratne et al., 2018). $LRM_{reg}$ focuses more on counteracting warming in specific areas than on the global average. In particular, it focuses on counteracting hot extremes in densely populated and agriculturally important regions. Because of its more regional focus, and a general reliance on established techniques, $LRM_{reg}$ avoids key problems that are associated with global SRM, such as detrimental impacts and regionally excessive responses, also known as overshoots. As stated by Seneviratne et al. (2018): "Land radiative management is not aiming to reduce the global temperature and so allows some regional limitation of the effects."

Some $LRM_{reg}$ proposals suggest making rooftops and walls of buildings brighter and more reflective, to help counteract the impact of heatwaves in cities. This may be done by painting them white; in some countries, for example around the Mediterranean Sea, white coatings on buildings have been the norm for a long time already (Figure 5.5). But similar or stronger effects can also be achieved with new reflective materials or coatings on roofs and walls elsewhere. Albedo increases in urban regions may have locally desirable effects, but globally these measures will only reflect an extra $0.01–0.05$ $W/m^2$ back into space (Lenton and Vaughan, 2009). At the very least, however, these measures would result in cooler buildings that require less air conditioning and thus operate at lower energy-consumption levels. Extrapolating the albedo increase measures from just urban areas to all areas involved in human settlement, the same authors find only an increase to $0.15$ $W/m^2$ of extra reflection.

**Figure 5.5** Whitewashed houses of Lindos, on the island of Rhodes, Greece.
*Source*: Saffron Blaze, CC BY 3.0, via Wikimedia Commons. https://commons.wikimedia.org/wiki/File:Lindos_Village_Overview.jpg

Other $LRM_{reg}$ approaches suggest the application of comparable techniques to crops. For example, gene manipulation could give plants a waxy, more reflective sheen, or farmers could simply select more reflective crops to replace less reflective crops. These approaches might decrease local and regional temperatures, with—again—a focus on heatwaves. It has been suggested that a 1°C cooling might be possible in temperate regions just by changing the reflectivity of crops (Ridgwell et al., 2009). Estimates of extra incoming sunlight reflection through cropland changes range between 0.14 and 0.35 $W/m^2$ (Lenton and Vaughan, 2009). The same study concludes that: "variegated plants, light shrubs, or bioengineered grasses and shrubs could be used to increase the albedo of grassland, open shrubland and savanna" to drive 0.5–0.6 $W/m^2$ of extra reflection, resulting in cooling.

Agriculture can make a further contribution through no-till farming practices. These are conducive to surface reflectivity increases after harvest, because they retain crop residues that have a higher albedo than bare ground (Seneviratne et al., 2018). The authors note that the impact is maximal for crops with high-reflectivity residues such as wheat. Because the greatest impacts take place after harvest, the maximum reflectivity gain could coincide with the times of most frequent heatwaves, for instance in August for winter wheat in Europe. No-till management in the case of summer crops modifies albedo in fall and potentially spring, but not in the hot summer months (the growing season).

However, things are never as simple as they seem to be at first glance. Recent work has found that earthworm-driven production of $N_2O$ and $CO_2$ emissions under no-tillage conditions "increase the global warming potential to the same level as in conventionally tilled [conditions], and are likely to offset most reductions in radiative forcing achieved by no-tillage management" (Lubbers et al., 2015). This result links in with reports based on assessments of soil carbon accumulation that no-tillage practices may have limited potential for mitigating climate change (Powlson et al., 2014). These potential complications perfectly illustrate the importance of whole-system assessments.

A pretty wild proposal has been made to cover deserts with reflective sheeting, which would consist of a white polyethylene top surface and an aluminum bottom surface (Gaskill, 2004). Covering 2% of Earth's surface this way can theoretically result in about 1.45 $W/m^2$ of extra reflection (Lenton and Vaughan, 2009), albeit at the not inconsequential cost of desert ecosystems. It remains to be seen how effective this dramatic method would be in the longer term, in view of dust settling on the sheeting and reducing its reflectivity.

Finally, albedo changes can be made with reforestation and deforestation. In tropical areas, reforestation of cleared lands has some cooling effect. But by far the biggest impact on reflectivity of the planetary surface from tree-cover changes is found in high-latitude and high-altitude forests, where deforestation exposes snow and thus increases the landscape reflectivity. Jiao et al. (2017) calculated that, if all of the world's forest cover measured in the year 2000 were deforested and replaced by crops, the planet's reflectivity would increase by about 0.8 $W/m^2$, which would cause cooling. Of course, this ignores the carbon-cycle impact of death and decay of all that forest cover, pumping carbon into the climate system in the form of methane and $CO_2$. We're here considering only the direct reflectivity, or albedo, change. In the opposite sense, the albedo-only impact of a complete global reforestation of present-day croplands toward the forest cover of the year 1700 amounts to about 0.2 $W/m^2$ reduction in reflectivity, which would produce warming.

But such widespread forest expansion would at the same time remove a large amount of $CO_2$ from the climate system. The net effects on climate depend on the extent to which one process compensates for the other, and it is hard to predict a priori how that will turn out.

## 5.6.  Reforestation

It is perhaps unexpected to see reforestation here under SRM methods, but it is important in that context, in addition to its potential role as a NET (section 4.2.1). This is because reforestation drives increased moisture flux through the vegetation into the atmosphere by the process of evapotranspiration. It thus causes increased cloudiness and rainfall, as well as land-surface cooling through both reflection of ISWR by clouds and heat used for evaporation. The latter effect will be familiar to most people from generally cooler conditions in the shade of a tree-covered park than in the shade of a nearby building or under an umbrella. Thus, large-scale reforestation results in considerable cooling as well as improved water cycling in the landscape, carbon removal from the atmosphere, and improved biodiversity. The opposite, further deforestation, causes exactly the opposite effects. Given that reforestation is not commonly ranked under the strict SRM approaches, this won't be done here either in the further SRM evaluation. But I have specifically highlighted it because it offers an important holistic approach of climate change management and alleviation, including SRM.

## 5.7.  Synthesis, Including Downsides and Challenges

SRM has long been viewed as an unnecessary and too risky step in dealing with climate change. But with unabated greenhouse gas emissions and increasing international pressure to limit climate change to within 1.5°C or 2°C, SRM has found its way back into the debating chambers. As we have seen, most approaches remain conceptual only, at best with limited model-based verification, or have been only indirectly tested in the field by roughly comparable processes (e.g., ship emissions). As a result, many questions remain about the long-term efficacy, feasibility, and likely detrimental environmental impacts of these approaches.

Some of the suggested SRM methods have strong theoretical potential for increasing reflectivity of incoming short-wave radiation and thus for driving cooling. But none, bar perhaps the iron salt aerosol approach, will directly help with reducing $CO_2$ levels, and thus with reducing global warming and ocean acidification. Neither can any SRM method generate the public health benefits that would accompany the phasing out of fossil fuel burning. Evidently, therefore, SRM methods cannot be considered in isolation from emissions reduction and avoidance, and negative emissions initiatives. If implemented, SRM should operate alongside those initiatives, so as to avoid continued build-up of dangerous unbalances in the oceans and climate.

Similar to the synthesis index that I presented for NETs, a synthesis index can be composed for SRM approaches (Figure 5.6). This exercise reveals that one SRM method with high potential, space mirror(s), falls short because of a very low TRL score, despite a reasonably low environmental impact score. If space mirror(s) were pursued, a very steep learning curve would need to be overcome before implementation would be possible at any scale, let alone at full scale. Two other methods come to the fore as most promising, namely stratospheric sulfate aerosol injection and marine cloud brightening. Both have their own drawbacks and challenges, but still come out high in the index scores because of relatively high TRL values and middling impact scores.

I emphasize that a simple index such as the one presented here cannot represent all relevant dimensions. Some of the lower-scoring methods—such as Arctic sea-ice thickening, human settlement albedo, and cropland albedo—have low scores only when considering their radiative influences on a global scale. Yet they are potentially very important on regional scales. After all, these methods are specifically designed to target regional problems, such as reducing the risk of Arctic warming and associated climate tipping points, making urban areas more habitable, and protecting crops and thus food supply against devastating heat waves.

In terms of specific drawbacks and challenges, all measures that block part of the incoming short-wave radiation from the Sun might reduce the amount of photosynthesis by plants and algae, and thus the amount of $CO_2$ uptake from the atmosphere and ocean. If this reduced $CO_2$ uptake is not perfectly offset by reduced $CO_2$ release during respiration and decomposition under cooler conditions, then more $CO_2$ will remain in the active climate system.

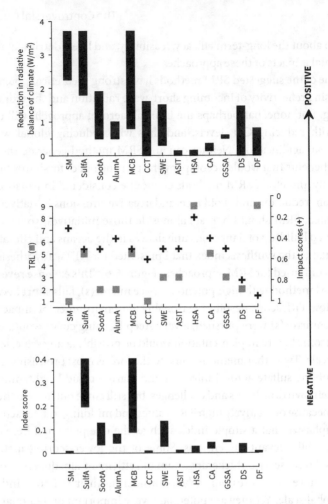

**Figure 5.6** Synthesis for the various SRMs. Top: minimum to maximum values for the additional reflection of incoming short-wave radiation. Middle: TRL and rough detrimental environmental impact scores between 0 and 1 (author's estimates). Bottom: index scores. SM: space mirror(s); SulfA: sulfate aerosol; SootA: soot aerosol; AlumA: aluminum aerosol; MCB: marine cloud brightening; CCT: cirrus cloud thinning; SWE: ships' wake extension; ASIT: Arctic sea-ice thickening; HSA: human settlement albedo; CA: cropland albedo; GSSA: grassland, shrubland, and savanna albedo; DS: desert sheets; DF: deforestation. Also indicated is the direction of impact for the various attributes: positive means best reflection, highest TRL, and least environmental impact, and negative vice versa. For input data, sources, and methodology behind this diagram, see Table 5.1.

This would result in warmer conditions than would exist if the $CO_2$ were removed by photosynthesis. In a complex, interlinked system, things are never easy and straightforward. Very careful assessment and continuous monitoring will be essential if we are to avoid implementing potentially expensive solutions to climate change, only to see them backfire immediately.

Similarly, all measures that block part of the incoming short-wave radiation will drive climate responses that differ both spatially and temporally from those driven by greenhouse gases that block part of the outgoing long-wave radiation. Changes in precipitation patterns are of particular concern because they can cause regionally devastating droughts, floods, and crop failure. Also, equator-to-pole temperature gradients may change, causing shifts in the mid-latitude storm track (MacMartin et al., 2018). And all SRM methods carry a risk of rapid warming if the SRM efforts are stopped for whatever reason, which may be challenging for ecosystems and human society alike.

With sulfate aerosols, concerns have been raised that the sulfuric acid will come down after a few years, bringing us straight back into the environmental problem of acid rain. In the 1970s and 1980s, acid rain caused by $SO_2$ (and $NO_x$) emissions from fossil-fuel burning for power generation and transport, and from animal-rearing in agriculture, became a matter of urgent environmental concern. Fortunately, its impacts could be curtailed by international actions to limit such emissions in the 1990s and 2000s. It would be imprudent to risk a return of this environmental disaster. But, note that the sulfate aerosol injection quantities are much smaller than the emissions that caused the acid rain problem. Hence, acid rain is not likely to become a major problem with sulfate aerosol injection. A bigger risk associated with this method is stratospheric ozone depletion that allows increased penetration of harmful UV light to the Earth's surface, where it damages the DNA in cells and causes cancer.

The impacts of marine cloud brightening are very localized—one can only make clouds where relative humidity is high already—but potentially large. The resultant changes in regional gradients may cause unfavorable weather patterns, upset ocean circulation, and severely stress local to regional ecosystems. Even if one succeeded at entirely evenly distributed cloud seeding over the oceans, this would only minimize such impacts and not eliminate them. Also, as we have seen before, any SRM-driven reduction in carbon sinks such as the Amazon rain forest—either because of reduced photosynthesis or because of precipitation changes—would reduce the influences of these sinks toward lowering global temperatures. If the SRM were then terminated, for whatever reason, the reduced carbon sinks can cause temperatures to overshoot to higher levels than those reached in the absence of SRM.

SRM methods that are more regionally focused, either deliberately so or by accident (as with the northern hemisphere focus of the ship wake extension method) are not free of challenges. Notably, there are ethical issues with the implementation of some of these methods. For example, there are issues around the right to interfere with crop-growing practices and regional agricultural management, which affect crop productivity and therefore food production. Potential benefits of the SRM need to be carefully weighed against such potential negative side effects.

The risk of weaponization of climate manipulation methods may sound too much like science fiction to really register with many people. Yet, weaponization of weather manipulation has been undertaken already. The United States ran a weather modification program known as Operation Popeye from March 20, 1967, until July 5, 1972, during the Vietnam War. Through cloud seeding, it aimed to extend the monsoon season, so as to limit troop and supply movements along the Ho Chi Minh Trail. SRM might be similarly deployed, for example with the objective of creating drought and crop failures over enemy territories. The UN Convention on the Prohibition of Military or Any Other Hostile Use of Environmental Modification Techniques (United Nations, 1978) only reduces the risk; it does not eliminate it. The prospect of targeted climate manipulation by SRM is generally seen as unlikely, but it cannot be completely ruled out for nations with enough money and infrastructure.

As it stands, resolving the issue of SRM governance is even more pressing than considering the risk of potential weaponization. Here, governance stands for goal-oriented, sustained, and explicit use of authority to influence behavior (Reynolds, 2019). That document breaks down governance issues around SRM into many bite-sized morsels; I am happy to refer the reader to it for details and will here only briefly touch upon a fraction of the international aspects involved.

Before the international community is ready for SRM implementation or large-scale testing, clarity will be needed about who will be responsible for implementing and regulating SRM applications. It is easy to think that these responsibilities will lie with the major nations currently

**Table 5.1** Synthesis for the various SRMs; basis for the numbers presented in Figures 5.6.

| | TRL | Radiative change Best (W/m2) | Radiative change Worst (W/m2) | Mean Env. Hazard (MEH) | Mean Environmental Hazard scores subjectively based on: |
|---|---|---|---|---|---|
| SM | 1 | 3.70 | 2.25 | 0.30 | Little intrusion, but potential impacts on photosynthesis and hydrological cycle; deployment risks |
| SulfA | 5 | 3.70 | 1.80 | 0.50 | Regional weather upsets (flood, drought); feedback processes; deployment issues |
| SootA | 2 | 1.90 | 0.50 | 0.50 | Regional weather upsets (flood, drought); feedback processes; deployment issues |
| AlumA | 2 | 1.00 | 0.50 | 0.40 | Regional weather upsets (flood, drought); feedback processes; deployment issues |
| MCB | 5 | 3.70 | 0.80 | 0.50 | Strongly patterned impacts on weather (flood, drought); feedback processes; deployment issues |
| CCT | 1 | 1.60 | 0.00 | 0.60 | As yet, much too poorly understood |
| SWE | 3 | 1.50 | 0.30 | 0.50 | Environmental impacts of surfactants; air-sea gas-exchange feedbacks |
| ASIT | 3 | 0.14 | 0.00 | 0.50 | Intrusive infrastructure in vulnerable environments; risk of pollution when equipment fails or during maintenance |
| HSA | 9 | 0.15 | 0.00 | 0.20 | Little impact, but effects mostly felt on local scales only |
| CA | 5 | 0.35 | 0.14 | 0.40 | Impacts on type of crops possible; possible adverse feedbacks especially with no-tillage approaches |
| GSSA | 3 | 0.60 | 0.50 | 0.50 | Impacts on existing ecosystems and biodiversity; possible adverse feedbacks |
| DS | 2 | 1.45 | 0.00 | 0.80 | Spatially extensive destruction of existing ecosystems |
| DF | 9 | 0.80 | 0.00 | 0.95 | Profound impacts on existing ecosystems and biodiversity; likely adverse feedbacks |

*Continued*

**Table 5.1** *Continued*

| | TRL 1-(1/TRL) | Radiative change Best (value/3.7) | Radiative change Worst (value/3.7) | 1-MEH | Index Best | Index Worst |
|---|---|---|---|---|---|---|
| SM | 0.00 | 1.00 | 0.61 | 0.70 | 0.00 | 0.01* |
| SulfA | 0.80 | 1.00 | 0.49 | 0.50 | 0.20 | 0.40 |
| SootA | 0.50 | 0.51 | 0.14 | 0.50 | 0.03 | 0.13 |
| AlumA | 0.50 | 0.27 | 0.14 | 0.60 | 0.04 | 0.08 |
| MCB | 0.80 | 1.00 | 0.22 | 0.50 | 0.09 | 0.40 |
| CCT | 0.00 | 0.43 | 0.00 | 0.40 | 0.00 | 0.01* |
| SWE | 0.67 | 0.41 | 0.08 | 0.50 | 0.03 | 0.14 |
| ASIT | 0.67 | 0.04 | 0.00 | 0.50 | 0.00 | 0.01 |
| HSA | 0.89 | 0.04 | 0.00 | 0.80 | 0.00 | 0.03 |
| CA | 0.80 | 0.09 | 0.04 | 0.60 | 0.02 | 0.05 |
| GSSA | 0.67 | 0.16 | 0.14 | 0.50 | 0.05 | 0.05 |
| DS | 0.50 | 0.39 | 0.00 | 0.20 | 0.00 | 0.04 |
| DF | 0.89 | 0.22 | 0.00 | 0.05 | 0.00 | 0.01 |

*Manually assigned maximum to achieve visibility in Figure 5.6.

*Sources:* Top panel synthesis of results from the impact of the Pinatubo eruption, Jones et al., (2009), Latham et al. (2008), Lenton and Vaughan (2009), Rasch et al. (2009), Shepherd et al. (2009), Muri et al. (2014), Crook et al. (2016), Desch et al. (2017), Jiao et al. (2017), and Lawrence et al. (2018).

*Notes:* Scores under "Mean Environmental Impact (MEH)" reflect author's subjective judgement, based predominantly on the qualitative arguments listed in the final column of the top panel. Index totals (plotted in Figure 5.6) calculated in the bottom panel using $T_{RL}$, which is $1-(1/TRL)$ so that the index gives greater weight to improvements between early TRL stages when innovation and creativity dominate advances, relative to improvement between later TRL stages when TRL largely depends on money invested; $R_p$ the radiative change potential normalized to a range between 0 and 1 by dividing all values in the top panel by the maximum value considered (3.7 W/m²; here considered the maximum value to avoid over-optimistic estimates based on extreme scenarios); and $I_{MP}$ the complement of the MEH core (i.e., 1–MEH). A complement value is used for $I_{MP}$ to ensure that the directions of all index components are harmonized to go from 0 = worst to 1 = best; in effect this inverts the reversed axis shown in the middle panel of Figure 5.6. The bottom panel's index scores are calculated in a simple linear, equally weighted manner: $I = R_p \times T_{RL} \times I_{MP}$

undertaking SRM research. But that would more or less exclude developing nations, which currently experience disproportionate impacts from climate change that was mainly caused by the major industrialized nations. The need to address this properly provides a powerful argument for deep involvement of developing nations in SRM research, development, evaluation, and implementation (e.g., Solar Radiation Management Governance Initiative, 2020). Deep involvement of developing nations is also important because "the belief that countries deploying SRM cause extreme weather events would create fertile ground for international conflict. Furthermore, it is unclear what the exact target global mean temperature should be, and who could legitimately make that decision, given the regionally uneven distribution of risks and benefits" (Council on Foreign Relations, 2019). Note here that the focus is not weather manipulation, but on manipulation of the overall climate state, which may affect—even unwittingly—patterns of weather extremes.

Yet, there are positive findings too; SRM generally may be conducive to reduction in income inequality between nations (Harding et al., 2020). Overall, the discussion must therefore be inclusive and transparent, and any implementation must be grounded in broad international agreement. The discussions around implementation of national pledges in accordance with the Paris Climate Agreement present interesting insight into the difficulties that will arise in the search for operational agreements with respect to SRM implementation. SRM has a certain potential in helping to get climate change under control, so such agreements will be vitally important. It's not impossible, but it won't be quick and easy; hence, it's best to start this process sooner rather than later.

# Key Sources and Further Reading

Alterskjær, K., and Kristjánsson, J.E. The sign of the radiative forcing from marine cloud brightening depends on both particle size and injection amount. *Geophysical Research Letters, 40*, 210–215, 2013. https://agupubs.onlinelibrary.wiley.com/doi/epdf/10.1029/2012GL054286

Climate Game Changers. *Climate restoration with Iron Salt Aerosol (ISA).* Accessed February 6, 2020. https://climategamechangers.org/game-changers/iron-salt-aerosol/

Council on Foreign Relations. *The anticipatory governance of Solar Radiation Management.* July 2, 2019. https://www.cfr.org/report/anticipatory-governance-solar-radiation-management

Crook, J.A., et al. Can increasing albedo of existing ship wakes reduce climate change? *Journal of Geophysical Research, Atmospheres, 121,* 1549–1558, 2016. https://agupubs.onlinelibrary.wiley.com/doi/epdf/10.1002/2015JD024201

Curry, C.L., et al., A multimodel examination of climate extremes in an idealized geoengineering experiment. *Journal of Geophysical Research, Atmosphere, 119,* 3900–3923, 2014. https://agupubs.onlinelibrary.wiley.com/doi/pdfdirect/10.1002/2013JD020648

Desch, S.J. et al. Arctic ice management. *Earth's Future, 5,* 107–127, 2017. https://doi.org/10.1002/2016EF000410

Dunne, D. Explainer: six ideas to limit global warming with solar geoengineering. *CarbonBrief: Explainer,* May 9, 2018. https://www.carbonbrief.org/explainer-six-ideas-to-limit-global-warming-with-solar-geoengineering

Evans, J.G.R., et al. Can oceanic foams limit global warming? *Climate Research, 42,* 155–160, 2010. https://pdfs.semanticscholar.org/9efe/53da9911cc80de2333184bbb13933d366926.pdf?_ga=2.268224141.228269493.1581060533-1519169105.1580084013

Eyring, V., et al. Transport impacts on atmosphere and climate: shipping. *Atmospheric Environment, 44,* 4735–4771, 2010. https://www.geos.ed.ac.uk/~dstevens/publications/eyring_ae09.pdf

Field, L., et al. Increasing Arctic sea ice albedo using localized reversible geoengineering. *Earth's Future, 6,* 882–901, 2018. https://agupubs.onlinelibrary.wiley.com/doi/epdf/10.1029/2018EF000820

Fuglestvedt, J., et al. Shipping emissions: from cooling to warming of climate—and reducing impacts on health. *Environmental Science and Technology, 43,* 9057–9062, 2009. https://pubs.acs.org/doi/pdf/10.1021/es901944r

Gaskill, A. *Summary of meeting with US DOE to discuss geoengineering options to prevent abrupt and long-term climate change.* July 29, 2004. http://www.homepages.ed.ac.uk/shs/Climatechange/Geo-politics/Gaskill%20DOE.pdf

Harding, A.R., et al. Climate econometric models indicate solar geoengineering would reduce inter-country income inequality. *Nature Communications, 11,* 227, 2020. https://www.nature.com/articles/s41467-019-13957-x

Hoppe, K. Mt. Pinatubo's cloud shades global climate. *Science News,* July 18, 1992. https://www.thefreelibrary.com/_/print/PrintArticle.aspx?id=12467057

ISA Australia. *Iron Salt Aerosol.* 2019. http://ironsaltaerosol.com/home

Jiao, T., et al. Global climate forcing from albedo change caused by large-scale deforestation and reforestation: quantification and attribution of geographic variation. *Climatic Change, 142,* 463–476, 2017. https://link.springer.com/article/10.1007/s10584-017-1962-8

Jones, A., et al. Climate impacts of geoengineering marine stratocumulus clouds. *Journal of Geophysical Research, Atmospheres, 114,* D10106, 2009. https://agupubs.onlinelibrary.wiley.com/doi/epdf/10.1029/2008JD011450

Jones, A., et al. A comparison of the climate impacts of geoengineering by stratospheric $SO_2$ injection and by brightening of marine stratocumulus cloud. *Atmospheric Science Letters, 12,* 176–183, 2011. https://rmets.onlinelibrary.wiley.com/doi/epdf/10.1002/asl.291

Jones, A.C., et al. Impacts of hemispheric solar geoengineering on tropical cyclone frequency. *Nature Communications, 8,* 1382, 2017. https://www.nature.com/articles/s41467-017-01606-0

Kaufman, R. Could space mirrors stop global warming? *LiveScience,* August 8, 2012. https://www.livescience.com/22202-space-mirrors-global-warming.html

Keith, D.W., et al. Stratospheric solar geoengineering without ozone loss. *Proceedings of the National Academy of Sciences of the USA, 113,* 14910–14914, 2016. https://www.pnas.org/content/pnas/113/52/14910.full.pdf

Latham, J., et al. Global temperature stabilization via controlled albedo enhancement of low-level maritime clouds. *Philosophical Transactions of the Royal Society A, 366,* 3969–3987, 2008. https://royalsocietypublishing.org/doi/pdf/10.1098/rsta.2008.0137

Latham, J., et al. Marine cloud brightening: regional applications. *Philosophical Transactions of the Royal Society A, 372,* 20140053, 2014. https://royalsocietypublishing.org/doi/pdf/10.1098/rsta.2014.0053

Lawrence, M.G., et al. 2018. As in Chapter 4.

Lenton, T.M., and Vaughan, N.E. The radiative forcing potential of different climate geoengineering options. *Atmospheric Chemistry and Physics, 9,* 5539–5561, 2009. https://www.atmos-chem-phys.net/9/5539/2009/acp-9-5539-2009.pdf

Lubbers, I.M., et al. Reduced greenhouse gas mitigation potential of no-tillage soils through earthworm activity. *Scientific Reports, 5,* 13787, 2015. https://www.nature.com/articles/srep13787

MacMartin, D.G., et al. Solar geoengineering as part of an overall strategy for meeting the 1.5°C Paris target. *Philosophical Transactions of the Royal Society A, 376,* 20160454, 2018. https://royalsocietypublishing.org/doi/pdf/10.1098/rsta.2016.0454

Matsui, H. Anthropogenic combustion iron as a complex climate forcer. *Nature Communications*, *9*, 1593, 2018. https://www.ncbi.nlm.nih.gov/pmc/articles/PMC5913250/

McKie, R. Could a £400bn plan to refreeze the Arctic before the ice melts really work? *The Guardian*, February 12, 2017. https://www.theguardian.com/world/2017/feb/12/plan-to-refreeze-arctic-before-ice-goes-for-good-climate-change

Milman, O. Could sprinkling sand save the Arctic's shrinking sea ice? *The Guardian*, April 23, 2018. https://www.theguardian.com/world/2018/apr/23/sprinkling-sand-save-arctic-shrinking-sea-ice

Ming, T., et al. 2014. As in Chapter 4.

Moriyama, R., et al. The cost of stratospheric climate engineering revisited. *Mitigation and Adaptation Strategies for Global Change*, *22*, 1207–1228, 2016. https://link.springer.com/article/10.1007%2Fs11027-016-9723-y

Muri, H., et al. The climatic effects of modifying cirrus clouds in a climate engineering framework. *Journal of Geophysical Research, Atmospheres*, *119*, 4174–4191, 2014. https://www.duo.uio.no/bitstream/handle/10852/58745/Muri_jgrd51304.pdf?sequence=1

National Research Council. *Climate intervention: reflecting sunlight to cool Earth.* The National Academies Press, Washington, DC, 260 pp., 2015. https://doi.org/10.17226/18988

Oeste, F.D., et al. Climate engineering by mimicking natural dust climate control: the iron salt aerosol method. *Earth System Dynamics*, *8*, 1–54, 2017. https://www.earth-syst-dynam.net/8/1/2017/esd-8-1-2017.pdf

Pierce, J.R., et al. Efficient formation of stratospheric aerosol for climate engineering by emission of condensible vapor from aircraft. *Geophysical Research Letters*, *37*, L18805, 2010. https://agupubs.onlinelibrary.wiley.com/doi/pdf/10.1029/2010GL043975

Powlson, D.S., et al. Limited potential of no-till agriculture for climate change mitigation. *Nature Climate Change*, *4*, 678–683, 2014. https://www.researchgate.net/publication/273621217_Limited_potential_of_no-till_agriculture_for_climate_change_mitigation

Rasch, P.J., et al. Geoengineering by cloud seeding: influence on sea ice and climate system. *Environmental Research Letters*, *4*, 045112, 2009. https://www.researchgate.net/publication/231088399_Geoengineering_by_cloud_seeding_Influence_on_sea_ice_and_climate_system

Reynolds, J. Solar geoengineering to reduce climate change: a review of governance proposals. *Proceedings of the Royal Society A*, *475*, 20190255, 2019. https://royalsocietypublishing.org/doi/pdf/10.1098/rspa.2019.0255

Ridgwell, A., et al. Tackling regional climate change by leaf albedo bio-geoengineering. *Current Biology*, *19*, 146–150, 2009. https://www.sciencedirect.com/science/article/pii/S0960982208016801

Robock, A., et al. Benefits, risks, and costs of stratospheric geoengineering. *Geophysical Research Letters*, *36*, L19703, 2009. https://agupubs.onlinelibrary.wiley.com/doi/pdf/10.1029/2009GL039209

Salter, S., et al. Sea-going hardware for the cloud albedo method of reversing global warming. *Philosophical Transactions of the Royal Society A, 366*, 3989–4006, 2008. https://royalsocietypublishing.org/doi/10.1098/rsta.2008.0136

Seitz, R. Bright water: hydrosols, water conservation and climate change. *Climate Change*, *105*, 365–381, 2011. https://www.researchgate.net/publication/225164197_Bright_water_Hydrosols_water_conservation_and_climate_change

Self, S., et al. The atmospheric impact of the 1991 Mount Pinatubo eruption. In: Newhall, C.G. and Punongbayan, R.S. (eds.). *Fire and mud: eruptions and lahars of Mount Pinatubo, Philippines*, pp. 1–43, University of Washington Press, Seattle and London, 1996. https://pubs.usgs.gov/pinatubo/self/

Seneviratne, S.I., et al. Land radiative management as contributor to regional-scale climate adaptation and mitigation. *Nature Geoscience*, *11*, 88–96, 2018. https://www.nature.com/articles/s41561-017-0057-5

Shepherd, J.G., et al. 2009. As in Chapter 4.

Solar Radiation Management Governance Initiative. Accessed May 26, 2020. http://www.srmgi.org

Storelvmo, T., et al. Cirrus cloud seeding has potential to cool climate. *Geophysical Research Letters*, *40*, 178–182, 2013. https://agupubs.onlinelibrary.wiley.com/doi/epdf/10.1029/2012GL054201

Temple, J. We're about to kill a massive, accidental experiment in reducing global warming. *MIT Technology Review*, January 22, 2018. https://www.technologyreview.com/s/610007/were-about-to-kill-a-massive-accidental-experiment-in-halting-global-warming/

United Nations. Convention on the prohibition of military or any other hostile use of environmental modification techniques. *Treaty Series, 1108*, pp. 151 and depositary notification C.N.263,1978. https://treaties.un.org/pages/ViewDetails.aspx?src=TREATY&mtdsg_no=XXVI-1&chapter=26lang=en

Wood, R., and Ackerman, T.P. Defining success and limits of field experiments to test geoengineering by marine cloud brightening. *Climate Change, 121*, 459–472, 2013. https://link.springer.com/article/10.1007/s10584-013-0932-z

# 6
# The Inevitable One
## Impacts and Adaptation

Our discussion to this point has been focused on methods by which climate change might be slowed down, stabilized, or even reversed; we will now consider processes for which it is already too late. Whatever we do, these are going to continue to change for the foreseeable future because they are governed by slow responses or feedbacks in the climate system. They are like the proverbial freight train that, once in motion, takes a long time to slow down. The emissions pathway humanity follows over the next century or so will determine the peak values reached by these slow processes (like the final speed of the freight train) and, hence, the time it takes for the slow processes to keep increasing before they eventually stabilize or can be reversed again.

Two slow processes in the climate system are of particular concern. They are slow because the systems involved are massive; they have a lot of inertia. One of the slow processes is ocean warming. The other is sea-level rise caused by thermal expansion due to ocean warming and by reduction in the volume of land-based ice. These processes have only just begun adjusting to the climate forcing that humanity has caused over the last 150–200 years, and especially the last 70 years. Ocean warming is relatively rapid, taking "only" 100–200 years for the uppermost 700–1000 meters of the oceans. But full temperature adjustment in the deep ocean takes many additional centuries, up to more than a millennium. Ice-volume changes happen over timescales of centuries to millennia, especially where the large ice sheets of Greenland and Antarctica are concerned.

As a result of the inertia, ocean temperature and land-ice volume are currently out of equilibrium with the radiative forcing of climate, and both will continue to adjust for centuries or more, even if we achieved zero emissions today. In reality, we are not achieving zero emissions today, and the amount of additional climate forcing applied from today, and the rate of its change, will determine by how much these systems eventually will adjust over and above the adjustment to their current disequilibrium. Even if we achieved negative emissions to the point of completely reducing the climate forcing to zero by 2100, the slow responses would largely play out before the high-inertia systems even "realized" that climate forcing had been reduced; a bit like hitting the brakes on a freight train at full speed. At first nothing much happens, and then very gradually it starts to slow down; it takes a very long time to stop. And to further complicate matters, we will see that once ice sheets have become reduced in size, re-growth is hindered over timescales of many centuries by a number of feedback processes. Hence, increased sea levels will remain high for hundreds of years.

In the following sections of this chapter, we look at what these slow processes have in store for us—their impacts are now to a large extent inevitable, regardless of which future emissions scenario we will follow. The inevitability means that humanity will have to adapt to these impacts.

## 6.1. Ocean Warming

Oceans warm up because water absorbs incoming short-wave radiation (sunlight), which excites the molecules and thus causes an increase in temperature. The geothermal heat flux from below is negligible in comparison, on timescales of several centuries. Warmer water has a lower density than colder water, and warming from the surface therefore increases the vertical density gradient in the water column, which makes it more stably stratified, or layered. Increased stratification has two key consequences. First, it alters the mainly wind-driven circulation that mixes the upper few hundred meters of the ocean. Second, it slows the large-scale deep-water overturning circulation in the oceans.

The first consequence is self-explanatory, but the second one may need some explanation. Because of the sea-water density-reduction that results from warming, new deep water formed from warmer surface waters is less likely to attain the same densities as were reached during deep-water formation before the surface warming took place. Thus, newly formed, slightly warmer deep waters are less likely to sink to similar depths and displace older, slightly cooler deep waters. Here, I am simply assuming that salinity doesn't change; that would complicate the story in detail but not fundamentally.

Surface-water warming thus slows the deep-water overturning circulation in the oceans.

Deep-water replacement in the world ocean is a process that takes a millennium or more at the best of times, and increasing the vertical density gradient through surface warming makes it an even slower process. Hence, temperature adjustment of the oceans is more rapid for the surface several hundreds of meters that receive sunlight directly and are mixed by winds, than for deep waters that are isolated from direct sunlight and whose temperature adjustment relies on deep-water circulation. It will take centuries to achieve surface ocean adjustment and millennia to achieve a complete ocean temperature adjustment. Global average ocean warming by 2100 is projected to reach about 0.6–2.0°C in the top 100 meters, and about 0.3–0.6°C at about 1000 meters. The two values in each range represent the IPCC's most-mitigated scenario (RCP2.6) and the continued-business-as-usual emissions scenario (RCP8.5), respectively. All values are relative to the average temperatures for the period 1986–2005 (IPCC, 2014; Bahri et al., 2018).

Ocean warming has several important consequences. These include impacts on ecosystems, storms and hurricanes, and ocean oxygenation. Finally, ocean warming also affects thermal expansion and ice-shelf stability, and thus sea-level change. These impacts are discussed in the following.

\* \* \*

Ocean warming causes shifts in the habitat zones of fish species (Moustahfid et al., 2018). As a result, traditional fisheries stock migrate away, and new, invasive species may appear. However, species in the ecosystem that are fixed to the sea floor—such as mussels, oysters, and kelp—cannot migrate, and any fish species that depend on immobile food-web components for part of their life cycle cannot easily migrate either (Oxenford and Monnereau, 2018). Thus, ecosystem composition is drastically altered, with disappearance of certain predator/grazing species from the region and invasion of new predator/grazing species into the region. This, often combined with overfishing and ocean acidification, can create unbalances that result in algal proliferation; sea-urchin, starfish, jellyfish, or squid population explosion; and commercial fisheries changes. Some places may lose commercial resources, while other places may gain them (e.g., Leonard, 2010).

Coral-dominated reef ecosystems (areas with more than 30% coral cover) are particularly sensitive to bleaching and mortality in response to warming, and there is a long history of warnings that climate change will make this a terminal problem for reef ecosystems (Goreau and Hayes, 1994). Even lower greenhouse gas emission scenarios are likely to drive the elimination of most warm-water coral reefs by 2040–2050 (Moustahfid et al., 2018). The impacts on Earth's ecosystems will be tremendous, because coral reefs are home to about a third of all biodiversity in the oceans. There is also a global kelp decline, as we saw in section 4.3.2, including collapse of kelp ecosystems around Tasmania. A sobering synthesis of marine heat waves and their profound impacts may be found on MarineHeatWaves (2020).

Finally, ocean warming is a key player in Arctic sea-ice reduction. Even under the favorable IPCC scenario RCP2.6, Arctic sea-ice extent will see a total decline of about 50% from about 2040. Under conditions of continued business as usual (RCP8.5), the Arctic sea-ice decline continues until losses of more than 90% result by about 2070 (IPCC, 2019). This has severe ecosystem implications, including the disappearance of ice-dependent species and food webs and invasion of open-ocean species, similar to the dramatic changes that were seen in the Bering Sea and around Antarctica, as discussed in section 2.3.

These types of impacts from ocean warming are projected to challenge fisheries productivity in sensitive regions by the middle of this century (IPCC, 2014).

\* \* \*

Ocean warming causes changes in the frequency, intensity, and geographic distribution of major storms and hurricanes. Development of hurricanes— or typhoons, as they are called in the western Pacific—requires a sea surface layer warmer than 26.5°C and limited vertical wind shear; that is, wind speeds need to remain roughly the same at sea level and higher altitudes. These are important conditions because hurricanes are so-called vertical heat engines that require heat to drive evaporation at the sea surface and release it higher up in the atmosphere when condensation occurs. This heat transfer from the surface to high atmospheric levels sets up a powerful, self-sustaining, vertical air updraft, or convection. Hurricane development is impeded by vertical wind shear because this tears the vertical heat engine apart sideways before it gets going.

Under the right conditions, hurricanes develop from tropical storms that gain more and more intensity. As air is sucked faster and faster into the storm's intensifying low-pressure cell, the rotation of the Earth eventually deflects the airflow into a powerful circulation around a central eye: the infamous hurricane winds. Meanwhile, the intense evaporation-condensation cycle of hurricanes causes them to dump exceptional amounts of rainfall.

In addition to extreme winds and rainfall, hurricanes also cause abrupt and devastating changes in local sea levels, which are known as storm surges and which can reach heights of many meters. Storm surges are mainly driven by a hurricane's intense winds and rapidly change direction when a hurricane's eye moves over an affected region. Storm surges cause extensive coastal damage and can drive massive inland flooding as they penetrate up rivers and breach levees (Figure 6.1).

The past century's ocean warming trend has accelerated over the past three decades and is expected to cause a change to fewer but more intense tropical cyclones and hurricanes (Emanuel, 2008). As climate warming proceeds, the atmospheric evaporation-condensation-precipitation cycle will be further intensified. This will not only increase the energy and thus the intensity of hurricanes, it will also enhance rainfall from tropical cyclone and

**Figure 6.1** Aftermath of the storm surge of hurricane Katrina in New Orleans, 2005.

*Source*: C.G. Colman. https://pixabay.com/photos/hurricane-katrina-flooding-180538/

hurricane events (Emanuel, 2017). The latter study furthermore projects a likely increase in the proportion of hurricanes that undergo rapid intensity increase just before landfall, which makes them extra dangerous. In all, we may expect a major rise in flooding potential in response to increases in both storm surges and rainfall amounts.

The regions where sea surface temperatures exceed 26.5°C are not only going to get hotter, they are also expanding poleward into areas where such high sea surface temperatures have not existed before. This poleward expansion has progressed by about 60 km per decade over the past 30 years (Kossin et al., 2014). Some modeling work has predicted that even western Europe will become exposed to hurricanes and extra-tropical cyclones (Haarsma et al., 2013). This chimes with hurricane Ophelia's track onto Ireland and the UK on October 16, 2017, with impacts that were compounded by the immediately subsequent extra-tropical cyclone Brian on October 21, 2017. Brian itself resulted from a merger of atmospheric disturbances related to former hurricanes Maria and Lee. While this does not prove a climate-change relationship—violent extra-tropical cyclones have hit Britain and Ireland before, notably in 1987—it's striking that a true hurricane like Ophelia made an unusual northeastward excursion along a track that agrees with those predicted in the Haarsma et al. (2013) model for a warming world. As Haarsma et al. concluded: "we anticipate an increase in severe storms of predominantly tropical origin reaching western Europe as part of 21st century global warming." Other regions may similarly become zones of hurricane activity.

* * *

Ocean warming also has severe implications for oxygen levels in the oceans. Gases exchange between surface water and the atmosphere, and warmer water can take up less of the atmospheric gases than cooler water. Therefore, warming of surface waters causes a reduction in the surface-water capacity to take up oxygen ($O_2$), as well as carbon dioxide ($CO_2$). Incidentally, ocean uptake of $CO_2$ limits the rise in atmospheric $CO_2$ concentrations, so that as the oceans warm up, a greater proportion of new emissions will stay in the atmosphere. In consequence, for constant emissions, atmospheric $CO_2$ levels and, hence, temperature will rise faster in a warmer world than in a cooler world. But let's return our focus to oxygen.

Oxygenation of sea waters is regulated by air:sea gas exchanges at the surface. Circulation then needs to mix that $O_2$ to greater depths. As water-column stratification increases in response to warming, changes develop in the patterns of wind-driven mixing of the upper few hundreds of meters. In some places, that mixing may become more intense. In many places, however, it weakens and subsurface oxygenation decreases. The deep sea, below on average about 700 meters, entirely relies for its $O_2$ supply on new deep-water formation at the surface.

In a warmer, more stratified ocean, the combination of less efficient deep-water formation and decreasing initial $O_2$ concentrations in warmer, newly formed deep water causes the ocean interior to become less well oxygenated. To make matters worse, increasing temperature causes an increase in the rates of microbial decomposition of organic matter. Thus, for equal amounts of sinking, dead organic matter, more $O_2$ will be used in the ocean interior for decomposition (and $CO_2$ released, driving acidification). This again causes important $O_2$ loss in the interior of warmer oceans. Overall, the open ocean has lost about 2% of its oxygen over the past 50 years (Breitburg et al., 2018).

Open-ocean Oxygen Minimum Zones, or OMZs, are found between upper depths of about 150–500 meters, and basal depths of about 1100–1500 meters. They should not be confused with regional so-called dead zones, which are associated with near-coastal eutrophication by nutrient-rich runoff from adjacent land. Instead, OMZs are permanent mid-depth features that are associated with broad, productive expanses of the ocean (Gilly et al., 2013). While OMZ development is a function of ocean circulation and biological productivity, there has been a remarkable trend of ocean oxygen decrease and OMZ expansion since the 1960s. Over that period, OMZs globally have expanded by an area about the size of the European Union, and the volume of water completely devoid of oxygen (anoxic) has more than quadrupled. These changes were most likely caused by ocean warming (Breitburg et al., 2018).

OMZ waters are more acidic than well-ventilated ocean waters because of their high content of $CO_2$ that was released during organic matter decomposition, which consumed the $O_2$. This acidification will intensify as OMZs expand and intensify and as more and more human-caused $CO_2$ emissions get absorbed in the ocean. Thus, OMZ expansion in warmer oceans will lead to a triple-whammy impact on ecology from temperature rise, de-oxygenation, and increasing water acidity. Expansion of OMZs, furthermore, leads to changes in microbial processes, which result in increased releases of the powerful greenhouse gas $N_2O$.

Expansion of OMZs, both spatially and with respect to depth, affects habitat ranges of organisms that thrive in their tops and in the overlying oxygenated waters. Some impacts may be detrimental, and some may be positive. For example, habitat compression above vertically expanding OMZs concentrates prey animals in narrower depth ranges closer to the surface where there is more light, both of which increase the foraging success of predators. Spatial OMZ extension drives migrations, similar to the poleward range shifts of marine species that have occurred in response to climate change and warming. Overall impacts on ecosystems and associated fisheries may turn out positive or negative, depending on the regional circumstances and the species of interest (Gilly et al., 2013).

In some places around the world, wind systems cause upwelling; especially in coastal upwelling centers at eastern oceanic margins, and—because of monsoon winds—in the northwestern Indian Ocean. Coastal upwelling pumps waters from about 200 meters depth to much shallower levels. Since 1999, low-oxygen water from an offshore OMZ has occasionally been pumped up to depths of less than 50 meters along the northwestern coast of the United States (Gilly et al., 2013). Notably, anoxic waters were brought up to very shallow levels close to the shoreline of central Oregon in 2006. This condition persisted for four months and caused widespread mortality of sea-floor dwelling invertebrates (Chan et al., 2008; Breitburg et al., 2018). Similar temporary shoaling of the top of the OMZ onto the shallow shelf, up to depths of only 25 meters, occurs at the Chilean coast, with equally profound impacts on fisheries. It's also known to happen off the Namibian coast. To make matters worse, oxygen depletion is exacerbated in almost all coastal regions of the world by eutrophication that is driven by increasing human-caused nutrient inputs from land.

\* \* \*

Ocean warming causes thermal expansion of sea water, which in turn causes sea-level rise. Although warming causes water to expand only fractionally, even a minute expansion will cause a notable sea-level rise because the vertical column of ocean water that is affected is hundreds to thousands of meters thick. Sea-water density decreases as sea-water volume increases when temperature increases above the $-1.8°C$ freezing point. In other words, salty water has a lower freezing point than fresh water, and lacks fresh water's exceptional behavior of maximum density and minimum volume at 4°C.

While water density decreases with increasing temperature, it increases with increasing salinity and increasing pressure (depth). The so-called equation of state is used to calculate the precise changes in the density/volume of water for given changes in temperature and salinity in the vertical. Across a series of models, the long-term sea-level rise commitment works out to 0.4 meters per degree Centigrade of uniform warming throughout the oceans (Levermann et al., 2013). For reasons explained earlier, such uniform warming is unlikely on timescales of multiple centuries as the deep sea won't have caught up. In consequence, the stated 0.4 meters is an upper estimate over such timescales because warming won't be completed in the deep sea; for moderate warming scenarios, the projected value is closer to 0.2 meters (e.g., Rahmstorf, 2012).

\* \* \*

Ocean warming also causes sea-level rise through destabilization of so-called ice shelves (Figure 6.2). Ice shelves are the marine terminations of ice streams, which are relatively fast-flowing "rivers" of ice within an ice sheet. Ice shelves consist of outflowing continental ice, not of sea ice. As an ice stream flows out into the sea, part of it lies grounded on the sea bed, but once the sea bed dips deeper than 90% of the thickness of the ice shelf, the ice stream floats and thus forms an ice shelf. Ice shelves can be thinned because of incursion of relatively warm sea water into the cavity between the sea bed

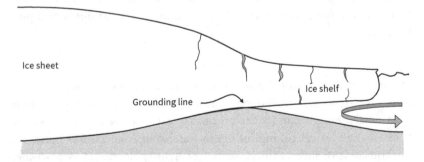

**Figure 6.2** Schematic of an ice shelf.
*Source*: Rohling, E.J. *The climate question: natural cycles, human impact, future outlook.* Oxford University Press, New York, 162 pp., 2019.

and the base of the ice shelf and/or because of surface melt related to atmospheric warming. Often, major portions break off in so-called calving events.

While Antarctic ice shelves have long been known to calve large icebergs on a regular basis, major ice-shelf collapses are known only from the last three decades. For example, the Larsen Ice Shelf has lost more than 75% of its former area in a series of rapid disintegrations, notably in 1995 and 2002 (NSIDC, 2020). Meanwhile, the Pine Island Ice Shelf has been thinning, large bergs have been breaking off, and major cracks have opened up that may herald a collapse event (Specktor, 2020). Following collapses, the feeding ice streams accelerate and bring masses of ice from land into the ocean. Around Greenland, ice shelves are less important because Greenland's glaciers terminate in narrow fjords that leave little room for ice shelves. But major calving-based retreat events still happen at its glacier edges.

Floating ice does not affect sea level, as it already directly displaces sea water. Similarly, grounded ice below sea level does not affect sea level when it directly replaces sea water, as is the case in depressions that have passages below sea level that connect them with the ocean; ice below sea level that is trapped in depressions without connections does not displace sea water. Regardless, ice-shelf thickness is a critical player in sea-level change (e.g., Hulbe, 2020). That is because once thinning happens, the ice shelf is lifted off the grounding line; the line that separates the grounded portion from the floating portion. In many cases, this grounding line is a ridge, and the bed dips away from it both in the seaward and the landward directions. The landward dip is caused by the massive ice sheet depressing the Earth's crust, as well as by the constant erosive action of the ice stream over millions of years. As the ice shelf is lifted up from the grounding line, there is a substantial reduction in bed friction, which in turn allows the supplying ice stream to accelerate. That acceleration brings land-based ice into the ocean, and thus drives sea-level rise. At the same time, relatively warm ocean water can penetrate deeper into the cavity and thus accelerate basal melting that further destabilizes the remaining ice shelf.

Total drainage of Antarctic Ice Sheet sectors sensitive to these processes might raise sea level by some 20 meters. The Greenland Ice Sheet (GrIS) lacks major ice shelves and is more sensitive to surface melt and outlet-glacier calving in its narrow fjords. Further discussion of the land-ice impacts on sea level follows in section 6.2.

## 6.2. Land-ice Reduction

Land ice exists in several forms. Most of it is stored in the great continental ice sheets of Greenland and Antarctica, with a smaller quantity in mountain ice caps and glaciers (Davies, 2019). In addition, there is ice in frozen ground, called permafrost.

Mountain ice caps and glaciers hold a mass equivalent to just over 0.4 meters of sea-level rise. Glaciers equilibrate to climatic changes on multidecadal to centennial timescales, while the typical timescales for the larger and flatter ice caps, like those in Alaska, are of the order of many centuries. The role of permafrost melting in sea-level rise remains uncertain, because it is not clear how much of the melt water may remain in the soil and how much of it may feed into rivers and eventually into the ocean.

Antarctica contains two ice sheets: the West Antarctic Ice Sheet (WAIS) and the East Antarctic Ice Sheet (EAIS). They are separated by the Transantarctic Mountain Range. WAIS holds an ice volume equivalent to about 4 meters of global sea-level rise and EAIS holds up to 54 meters. WAIS is grounded well below sea level, even after allowing for rebound (uplift) of the land following removal of the ice mass (Jamieson et al., 2014). Large portions of EAIS are also grounded below sea level; certainly initially, before rebound brings much of this area above sea level over thousands of years after the ice is removed. As we saw in the previous section, these portions amount to a volume equivalent to about 20 meters sea-level rise. The marine-based (grounded below sea level) portions of WAIS and also EAIS are of most concern regarding sea-level rise in the relatively near future. This is because dynamic interactions between the ocean and ice sheet in these regions (section 6.1) create conditions that are most conducive to rapid ice-mass loss and, thus, sea-level rise (Noble et al., 2020). It remains uncertain to what extent such marine ice-shelf instability processes, also known as MISI, may lead to collapse of large sectors of WAIS and/ or EAIS, but the process is increasingly included in sea-level assessments. Another potential process, marine ice-cliff instability, or MICI, is more contested among glaciologists, and has not been included in the final numbers of the most recent IPCC sea-level assessment (IPCC, 2019). Yet, observational evidence suggests that it is a very real process and that it may in future become a critically important mechanism, capable of causing very rapid sea-level rise (Parizek et al., 2019).

The ice-shelf and cliff processes apply also to projections of change in the GrIS—in fact, the observations of ice cliff instability actually derive from Helheim Glacier in Greenland. But they are less important for GrIS because it is in general more isolated from marine influences than WAIS and EAIS and instead suffers more from the impacts of atmospheric warming and meltwater-related processes atop and inside the ice sheet. GrIS holds a total mass equivalent to just over 7 meters sea-level rise. It has over the past several decades been showing enormous and accelerating surface melting and mass loss.

IPCC (2019) gave recent projections of mean sea-level rise from all sources combined, including ocean thermal expansion but without considering MICI. These amount to about 0.3 meters per century, up to 1.2 meters per century, when averaging projected changes over the period 2000–2300, and the two values represent IPCC scenarios RCP2.6 and RCP8.5, respectively. Another analysis, which explicitly includes MICI, suggests higher values of 0.5 and 3.9 meters per century, respectively (Kopp et al., 2017). As already mentioned, ice-shelf processes such as MISI and MICI are especially important for the Antarctic ice-volume contribution to sea-level rise.

For comparison, a recent study of the last interglacial—the last time global sea level rose up to 10 meters above the present level between about 130,000 and 118,000 years ago—reported three multi-century events of sea-level rise, with mean rates of rise of about 2.8, 2.3, and 0.6 meters per century (Rohling et al., 2019). Moreover, the study specifically attributed the two highest values to Antarctic ice-volume changes. The existence of such extreme rates, during a warm period well before humans became a player in climate change, demonstrates that the processes needed for such high rates are real and can be (and have been) activated, no matter whether they are called MICI or something else. The last interglacial was not much warmer than today, but heat redistribution processes within the oceans had caused water temperatures around Antarctica to be more similar to what we expect by 2100 in a world that is 1.5–2.0°C warmer than pre-industrial. So, if this is where the threshold lies for Antarctic instability and catastrophic sea-level rise, then there's a good reason not to push warming beyond the limits of the Paris Agreement. Recent work indicates that holding warming to a maximum of 2°C still risks sufficient melt-back of the Antarctic ice sheet alone to cause 2.5 meters of sea-level rise over several centuries, while 4°C warming would increase that to 6.5 meters. And this does not even count contributions from Greenland (Garbe et al., 2020; Harvey, 2020).

In sum, we have seen that, relative to year-2000 values, about 0.5 meters of sea-level rise by 2100 seems unavoidable even in the favorable RCP2.6 IPCC emissions scenario, which still allows warming to exceed the 1.5°C Paris Agreement target. In reality, we're nowhere near such an optimistic scenario; instead, the emissions scenario we're busily following is likely to result in sea-level rise of a meter or more by 2100.

Let's assume that we manage to get ourselves onto an emissions pathway that limits warming to the 1.5°C threshold of the Paris Agreement, through drastic emissions reduction and avoidance, and implementation of NETs. Even in that case, we will by 2100 have to contend with at least 0.35 meters of sea level rise above year-2000 levels (Nauels et al., 2019). This baseline of sea-level rise is driven by the delayed responses of high-inertia systems, notably ocean warming and ice-volume change, to the climate forcing that has been applied already since the industrial revolution. In technical jargon, we say that this sea-level rise component is committed to already, regardless of the emissions pathway that is followed from now on. And because the inertia of the slow-response systems is so great, the component of rise that is committed to already will continue to grow for centuries. For example, it will amount to roughly 0.8 meters by 2300. Because it is internal to the system in today's state, there is practically nothing we can do to stop this baseline rise from happening. It is to this baseline that the consequences of any further emissions from today are added.

## 6.3. Adaptation

Governance of climate change calls for action in both mitigation and remediation (Chapters 3–5), and adaptation. Mitigation and remediation require shifts in current behavior to end practices driving further climate change and to restore climate to safe levels. Adaptation implies adjustments of society to the impacts of a changing climate. Therefore, governance of adaptation requires planning to deal with expected impacts on both human activity and vital ecosystem services (Meadowcroft, 2009). Such planning requires knowledge of anticipated regional and local climate effects because climate change is not uniformly distributed over the planet. Instead, it describes a long-term average change in the climate state of the planet, which expresses itself with a wide range of variability on regional and local scales. For example, it may get very warm in certain regions, not much warmer in other regions, and even a little cooler in a few regions. Spatial contrasts are also expected in precipitation, evaporation, and wind systems.

It is important to emphasize at this point that many impacts of climate change should not be adapted to but rather avoided by strong climate mitigation and remediation. And, even better, the three response streams should be undertaken in a combined manner in an integrated response framework wherever possible. That would help with: (a) minimizing the impacts to which adaptation is needed by ensuring that impacts are not made any worse than they are already locked in to be; and (b) freeing up effort and resources for adaptation where it is essentially the only option for impacts that are already locked in because they arise from slow, delayed responses of systems with great inertia. The dominant responses, by far, in the latter category are ocean warming and land-ice volume reduction. This section takes a brief look at key parameters for adaptation to these two processes.

With respect to ocean warming, we have seen that the dominant impacts relate to temperature change itself and to circulation change and deoxygenation. These processes cause shifting habitat zones and unbalanced ecosystems because of predator or prey removal, mass mortality from anoxia, and collapse of crucial habitats such as coral reefs and kelp forests. In estuaries—the fresh-to-salty transition zones at river mouths that are hotspots for marine biodiversity and fisheries—warming, sea-level rise, and potential tidal changes cause enhanced salt-water incursion and deoxygenation. Along with grave impacts on natural habitats and biodiversity, such changes in the marine realm seriously threaten commercial fisheries, aquaculture, tourism, local communities that are highly dependent on fisheries, as well as specialized indigenous cultures. Moreover, shifts in species distributions and abundance are challenging (inter-)national ocean and fisheries governance aimed at safeguarding ecosystems and resource-sharing between fishing entities (IPCC, 2019). Clearly, ocean warming will through its variety of consequences create enormous, geographically variable, social and economic costs.

Given that committed change is inevitable even in dramatic mitigation scenarios, all we can do is soften the abruptness of the impacts through timely planning and implementation of adaptations. Strong climate mitigation and remediation are needed at the same time to avoid a worsened situation. The sharp rise in future population numbers, and the importance of fisheries and coastal communities for sustainably feeding vast numbers of people, make adaptation planning and implementation matters of great urgency. This calls for immediate programs to develop regional skills and capacity, along with allocation of broad structural support and funds to the regions involved. Such programs can be structured to simultaneously deal with mitigation and remediation, so that a next generation is trained with an integrated view of how to tackle the problem on all fronts.

With respect to sea-level rise, the unavoidable minimum rise of 0.35 meters by 2100 may not sound like much, but it will more than double the 0.3 meter rise that took place from 1800 to 2000. It will displace the many people who currently live below or just above the current high-water lines. Today, about 110 million people live on land that falls below the high tide mark. Because of sea-level rise, this number climbs to about 190 million by the year 2100 under low emissions (RCP2.6) or 630 million under high emissions (RCP8.5). Looking above the high tide mark, 1 billion people today live within the first 10 meters. Of these, 440 million live between 1 and 5 meters and 230 million within the first 1 meter (Kulp and Strauss, 2019). These coastal communities are severely threatened by sea-level rise, and especially by its interaction with spring tides, storm surges, and extreme wave conditions and rainfall events. Options for coastal protection to safeguard current assets and livelihoods range from hard engineering to natural solutions such as reefs, mangroves, and dune fields. The alternative to protection is relocation of people, infrastructure, and activities. Different combinations will apply in different places with different settings, conditions, and expected regional impacts.

An important consideration is that the increase in flooding potential from sea-level rise in many regions doesn't just come from the sea, but from the landward side as well. This is because sea-level rise lifts up rivers in their near-coastal flow regions, which increases the risk that levees are overtopped during major rainfall events, especially when these are superimposed upon high tides and storm surges. In consequence, protective infrastructure needs to be reinforced and increased not only along the coast itself, but also along the lower flow regions of rivers. Alternatively, mobile barriers may be built across river mouths, which can be closed to keep out the sea during storms and/or extreme tides. When such barriers are closed, massive pumping stations may need to pump the natural river flow past the barrier into the sea, or else the swelling river itself might still overtop the levees. Yet, in some low-lying developing nations with deltas that have massive, complicated coastlines with many river outlets, construction of prohibitively expensive large-scale coastal flooding protection often isn't economically and/or geographically feasible. Sea-level rise—especially in combination with storm surges and extreme tides and waves—will therefore cause increasingly frequent to virtually permanent flooding of areas and assets that are critical to habitation and livelihoods.

Another important consideration is that even the most extensive coastal defense projects cannot stop increasing saltwater penetration into groundwater driven by sea-level rise. This penetration poisons drinking water reserves and causes severely inhibiting conditions for agriculture and natural flora, across many tens of kilometers inland from the coast in low-lying regions.

The combination of flooding and saltwater incursion destroys living conditions and livelihoods. Bangladesh is a case in point. Today, salt-water incursion into soils and drinking water, and widespread flooding, affect more than 33 million people in that nation. This situation is set to worsen significantly, as the nation stands to lose a further 10% or more of its territory with a 0.5 meter sea-level rise, which will directly displace 15 million people and drive additional widespread intensification of salt-water incursion. Moreover, Bangladesh is severely affected by tropical storms and hurricanes that are expected to gain in intensity, exacerbating the flooding risks. Eventually, destroyed living conditions and livelihoods leave only the option of abandoning the region, which results in mass migration and associated humanitarian, health, and security issues (Glennon, 2017; McDonnell, 2019).

Again, integration of adaptation with mitigation and remediation efforts offers exciting opportunities. For example, particularly elegant solutions to climate-change impacts would combine (re-)building improved coastal protection and marine ecosystem resilience with use of renewable energy resources and NETs implementation. We have encountered some of this under blue carbon in section 4.3.2, but blue carbon initiatives largely aim to reinstate ecosystems to strengthen current coastlines, rather than build new strength against expected sea-level rise and storm intensification. In contrast, electrically stimulated limestone growth, called Biorock, has proven benefits for newly establishing, revitalizing, and further strengthening coral reefs and non-coral coastal breakwaters.

Biorock may be as strong as, or stronger than, concrete and continues to gain strength with time as long as electric stimulation is continued. Furthermore, Biorock formation may help to draw down $CO_2$ from the climate system, and Biorock-based coral reefs have demonstrated exceptional potential for recovery after major bleaching events, as long as electric stimulation remained active. Finally, Biorock structures can grow at around 20 mm/year, or faster when superimposed corals and oysters are taken into account. This offers real potential to grow coastal protection that keeps up with sea-level rise, in addition to creating conditions to grow back severely eroded beaches in months while regenerating reefs, seagrasses, biodiversity, and fisheries (Goreau and Trench, 2012; Global Coral Reef Alliance, 2020). The weak electrical stimulation underpinning Biorock has been found to be safe, but nevertheless deters shark feeding. The absence of apex predators may then upset ecosystem balance, which requires further assessment (Uchoa et al., 2017).

Finally, so-called transition management may be an important tool for guiding both adaptation to and mitigation of climate change (Meadowcroft, 2009). It identifies promising pathways for the evolution of key societal sectors—the energy system, agriculture, health care, and so on—and develops a portfolio of transition experiments to accelerate innovation and explore potential avenues for change. Such efforts can be coordinated by governments, based on an (inter-)national framework of agreed aims and objectives, and then supported by governments and carried out in collaboration with key societal stakeholders.

As critical stakeholders, local communities and indigenous cultures should be fully involved from the outset in the process of planning and implementation of region-specific adaptation measures (Ramos-Castillo et al., 2017). They can provide important local knowledge and understanding that should be accounted for; they can provide personnel needed for implementing, monitoring, and regulating the adaptation measures; and they are key to the social acceptability of adaptation measures. Hence, inclusive measures accepted by, and possibly economically boosting, local communities are more likely to become a long-term success. Integration of these efforts with those focused on mitigation and remediation will further smooth the pathway to a more sustainable future, on both regional and global scales.

# Key Sources and Further Reading

Bahri, T., et al. Chapter 1: Climate change and aquatic systems. In: Barange, M., et al. (eds.). *Impacts of climate change on fisheries and aquaculture: synthesis of current knowledge, adaptation and mitigation options. FAO Fisheries and Aquaculture Technical Paper 627*, pp. 1–17, Rome, FAO, 2018. http://www.fao.org/3/i9705en/i9705en.pdf

Breitburg, D., et al. Declining oxygen in the global ocean and coastal waters. *Science, 359*, eaam7240, 2018. https://science.sciencemag.org/content/359/6371/eaam7240

Chan, F., et al. Emergence of anoxia in the California current large marine ecosystem. *Science, 319*, 920, 2008. https://science.sciencemag.org/content/319/5865/920

Davies, B. *What is the global volume of land ice and how is it changing?* AntarcticGlaciers. org, January 19, 2019. http://www.antarcticglaciers.org/glaciers-and-climate/what-is-the-global-volume-of-land-ice-and-how-is-it-changing/

Emanuel, K. The hurricane–climate connection. *Bulletin of the American Meteorological Society*, May 1, 2008. https://journals.ametsoc.org/doi/pdf/10.1175/BAMS-89-5-Emanuel

Emanuel, K. Will global warming make hurricane forecasting more difficult? *Bulletin of the American Meteorological Society*, March 31, 2017. https://journals.ametsoc.org/doi/pdf/10.1175/BAMS-D-16-0134.1

Environmental Justice Foundation. *Climate displacement in Bangladesh*. Accessed March 28, 2020. https://ejfoundation.org/reports/climate-displacement-in-bangladesh

Garbe, J., et al. The hysteresis of the Antarctic Ice Sheet. *Nature, 585*, 538–544, 2020. https://www.nature.com/articles/s41586-020-2727-5

Gilly, W.F., et al. Oceanographic and biological effects of shoaling of the Oxygen Minimum Zone. *Annual Reviews of Marine Science, 5*, 393–420, 2013. https://web.stanford.edu/group/MicheliLab/pdf/oceanographicandbiologicaleffectsofshoaling.pdf

Glennon, R. The unfolding tragedy of climate change in Bangladesh. *Scientific American*, April 21, 2017. https://blogs.scientificamerican.com/guest-blog/the-unfolding-tragedy-of-climate-change-in-bangladesh/

Global Coral Reef Alliance. *Biorock™, Mineral Accretion Technology™, Seament™*. Accessed March 31, 2020. http://www.globalcoral.org/biorock-coral-reef-marine-habitat-restoration/

Goreau, T.J., and Hayes, R.L. As in Chapter 2.

Goreau, T.J., and Trench, R.K. *Innovative methods of marine ecosystem restoration.* CRC Press, 312 pp., 2012.

Gruber, N. Warming up, turning sour, losing breath: ocean biogeochemistry under global change. *Philosophical Transactions of the Royal Society A, 369*, 1980–1996, 2011. https://royalsocietypublishing.org/doi/pdf/10.1098/rsta.2011.0003

Geological Society of America. Tall ice-cliffs may trigger big calving events—and fast sea-level rise. *Science Daily*, March 22, 2019. https://www.sciencedaily.com/releases/2019/03/190322163342.htm

Haarsma, R.J., et al. More hurricanes to hit western Europe due to global warming. *Geophysical Research Letters, 40*, 1783–1788, 2013. https://agupubs.onlinelibrary.wiley.com/doi/pdf/10.1002/grl.50360

Harvey, F. Melting Antarctic ice will raise sea level by 2.5 metres—even if Paris climate goals are met, study finds. *The Guardian*, September 24, 2020. https://www.theguardian.com/environment/2020/sep/23/melting-antarctic-ice-will-raise-sea-level-by-25-metres-even-if-paris-climate-goals-are-met-study-finds

Hulbe, C. How close is the West Antarctic ice sheet to a "tipping point"? *CarbonBrief*, February 14, 2020. https://www.carbonbrief.org/guest-post-how-close- is-the-west-antarctic-ice-sheet-to-a-tipping-point

IPCC. *Climate change 2014: synthesis report. Contribution of Working Groups I, II and III to the Fifth Assessment Report on the Intergovernmental Panel on Climate Change.* Pachauri, R.K. and Meyer, L.A., (eds.). Geneva, Intergovernmental Panel on Climate Change, 151 pp., 2014. http://www.ipcc.ch/report/ar5/syr/

IPCC. *IPCC Special Report on the Ocean and Cryosphere in a Changing Climate.* Pörtner, H.-O., et al. (eds.), Intergovernmental Panel on Climate Change, 755 pp., 2019. https://www.ipcc.ch/site/assets/uploads/sites/3/2019/12/SROCC_FullReport_FINAL.pdf

Jamieson, S.S.R., et al. The glacial geomorphology of the Antarctic ice sheet bed. *Antarctic Science, 26*, 724–741, 2014. https://www.researchgate.net/publication/280564159

Kopp, R.E., et al. Evolving understanding of Antarctic ice-sheet physics and ambiguity in probabilistic sea-level projections. *Earth's Future, 5*, 1217–1233, 2017. https://agupubs.onlinelibrary.wiley.com/doi/pdf/10.1002/2017EF000663

Kossin, J.P., et al. The poleward migration of the location of tropical cyclone maximum intensity. *Nature, 509*, 349–352, 2014. https://www.nature.com/articles/nature13278

Kulp, S.A., and Strauss, B.H. New elevation data triple estimates of global vulnerability to sea-level rise and coastal flooding. *Nature Communications, 10*, 4844, 2019. https://www.nature.com/articles/s41467-019-12808-z

Leonard, K. Fisheries and climate change. *Asian Development Bank, Policy Brief SPSO Oct 2010/003*, October 10, 2010. https://think-asia.org/bitstream/handle/11540/531/spso-201003-fisheries-climate-change.pdf?sequence=1

Levermann, A., et al. The multimillennial sea-level commitment of global warming. *Proceedings of the National Academy of Sciences of the USA, 110*, 13745–13750, 2013. https://www.pnas.org/content/110/34/13745

Locke, S.F. How it works: protecting New Orleans with the world's largest flood pump. *Popular Science*, August 18, 2009. https://www.popsci.com/scitech/article/2009-08/saving-new-orleans-worlds-largest-water-pump/

Maitre, N., et al. *The employment impact of climate change adaptation. Input Document for the G20 Climate Sustainability Working Group.* International Labour Office, 46 pp., Geneva, 2018. https://www.ilo.org/wcmsp5/groups/public/---ed_emp/documents/publication/wcms_645572.pdf

MarineHeatWaves. *Marine Heat Waves international working group.* Accessed March 25, 2020. http://www.marineheatwaves.org

Meadowcroft, J. Climate change governance: background paper to the 2010 world development. Worldbank Policy Research Working Paper, 4941, 2009. http://documents.worldbank.org/curated/en/210731468332049368/pdf/WPS4941.pdf

McDonnell, T. Climate change creates a new migration crisis for Bangladesh. *National Geographic*, January 24, 2019. https://www.nationalgeographic.com/environment/2019/01/climate-change-drives-migration-crisis-in-bangladesh-from-dhaka-sundabans/

Moustahfid, H., et al. Chapter 12: Climate change impacts, vulnerabilities and adaptations: Western Indian Ocean marine fisheries. In: Barange, M. et al. (eds.). *Impacts of climate change on fisheries and aquaculture: synthesis of current knowledge, adaptation and mitigation options. FAO Fisheries and Aquaculture Technical Paper, 627*, pp. 251–279, Rome, FAO, 2018. http://www.fao.org/3/i9705en/i9705en.pdf

Nauels, A., et al. Attributing long-term sea-level rise to Paris Agreement emission pledges. *Proceedings of the National Academy of Sciences of the USA, 116*, 23487–23492, 2019. https://www.pnas.org/content/116/47/23487

Noble, T., et al. The sensitivity of the Antarctic Ice Sheet to a changing climate: past, present, and future. *Reviews of Geophysics*, 2020. https://agupubs.onlinelibrary.wiley.com/doi/abs/10.1029/2019RG000663

NSIDC. *Larsen Ice Shelf breakup events.* National Snow and Ice Data Center. Accessed March 23, 2020. https://nsidc.org/news/newsroom/larsen_B/index.html

Oxenford, H.A., and Monnereau, I. Chapter 9: Climate change impacts, vulnerabilities and adaptations: Western Central Atlantic marine fisheries. In: Barange, M., et al. (eds.). *Impacts of climate change on fisheries and aquaculture: synthesis of current knowledge, adaptation and mitigation options. FAO Fisheries and Aquaculture Technical Paper, 627*, pp. 185–206, Rome, FAO, 2018. http://www.fao.org/3/i9705en/i9705en.pdf

Parizek, B.R., et al. Ice-cliff failure via retrogressive slumping. *Geology, 47*, 449–452, 2019. https://pubs.geoscienceworld.org/gsa/geology/article/47/5/449/569567/Icecliff-failure-via-retrogressive-slumping

Rahmstorf, S. Modeling sea level rise. *Nature Education Knowledge, 3*, 4, 2012. https://www.nature.com/scitable/knowledge/library/modeling-sea-level-rise-25857988/

Ramos-Castillo, A., et al. Indigenous peoples, local communities and climate change mitigation. *Climatic Change, 140*, 1–4, 2017. https://link.springer.com/article/10.1007/s10584-016-1873-0

Rohling, E. Perfect storm threatens the world's reefs. *Cosmos Magazine*, January 24, 2018. https://cosmosmagazine.com/geoscience/perfect-storm-threatens-the-world-s-reefs

Rohling, E.J., et al. A geological perspective on potential future sea-level rise. *Scientific Reports, 3*, 3461, 2013. https://www.nature.com/articles/srep03461

Rohling, E.J., et al. Asynchronous Antarctic and Greenland ice-volume contributions to the last interglacial sea-level highstand. *Nature Communications, 10*, 5040, 2019. https://www.nature.com/articles/s41467-019-12874-3

Rugerri, A. How climate change will transform business and the work-force. *BBC Future Now*, July 10, 2017. https://www.bbc.com/future/article/20170705-how-climate-change-could-transform-the-work-force

Specktor, B. One of Antarctica's fastest-shrinking glaciers just lost an iceberg twice the size of Washington, D.C. *LiveScience*, February 11, 2020. https://www.livescience.com/pine-island-glacier-calving-retreat.html

Szczepanski, M., et al. Bangladesh: a country underwater, a culture on the move. Natural Resources Defense Council, September 13, 2018. https://www.nrdc.org/onearth/bangladesh-country-underwater-culture-move

Uchoa, M.P., et al. The effects of Biorock-associated electric fields on the Caribbean reef shark (*Carcharhinus perezi*) and the bull shark (*Carcharhinus leucas*). *Animal Biology, 67*, 191–208, 2017. https://brill.com/view/journals/ab/67/3-4/article-p191_1.xml

Wikipedia. *Fox Point hurricane barrier*. Accessed March 28, 2020. https://en.wikipedia.org/wiki/Fox_Point_Hurricane_Barrier

# 7
# The Behavioral Renaissance
## Re-forming Society

As we have seen, readjusting the radiative balance of climate to a societally acceptable level, which according to the Paris Climate Agreement lies at around 1.5°C to at most 2°C warming, will require a complex portfolio of measures to achieve (1) emissions reduction, (2) new emissions avoidance, and (3) greenhouse gas removal, along with (4) potential solar radiation management. Moreover, the proportional representation of these four different classes of measures will need to be flexible through time, in response to different needs, such as a high need for emissions reduction today that may decline with time as emissions approach zero (see Figure 1.1). Flexibility will also be needed in response to the emergence of new breakthroughs, challenges, cost limits, and economic and societal constraints. Key parameters regarding societal change are discussed in section 7.1, while section 7.2 looks more specifically at the roles of government, corporations, and consumers. Finally, section 7.3 evaluates routes for channeling discontent and litigation.

## 7.1. Parameters of Change

Most of humanity's current actions seem to be predicated on an underlying assumption that resources and space on Earth are near infinite and that we have an innate right to take or discard whatever we like. Simply blaming this on a capitalist drive for profit by a handful of people would skirt around the real underlying issue: we are all addicted to convenience, or close to joining in this addiction. As a result, consumerism is rife in society. In vast quantities, we reject and replace rather than reduce, repair, reuse, and recycle. While industry definitely facilitates and promotes this behavior, that's not what keeps it going; at its heart, it comes down to our own collective choice.

We have all seen the power of the public's collective choice once a realization takes hold that the environment, or resources, are finite. Once sentiments take over and people decide that the world [or nature] can't take any more of this, or that enough is enough, fundamental changes can be driven on relatively short time scales through widespread public pressure. A good example is the current global pressure to reduce production of single-use plastics. Other largely public-pressure-driven initiatives are developing toward waste reduction, including zero-waste and circular-economy startups. Most times, however, the public is not so uniformly convinced of the urgency of change, even when expert assessments indicate a pressing need for it. Effective implementation of change then relies heavily on international and/or national processes of regulation. Environmental examples include the ban on ozone-destroying compounds and the ban on lead-additives to fuels and paint. Action to limit climate change sits between these two extremes; public pressure for it is high in some places at some times and in some population groups but weak to non-existent in others. So, where does this leave us in terms of the potential for making a difference? Are we stuck?

La Branche (2020) summarizes people's responses to the need for behavioral change in three categories that are based on social (not economic) factors such as perceptions, beliefs, social representations, and practices. First are the so-called resisters, who make up about 20% of the population. They consider any effort to change an erosion of their comfort and individual rights. Second are the so-called motors, making up about 60% of the population. They see the changes as a positive avenue for putting their values into practice. Third are the 20% of so-called variables. They modify their practices according to how the requirements align with the needs of their life phase, such as having a second child, buying a home, retirement, and so on. In essence, the proportions of these categories remain roughly constant across society, given that they represent deeply rooted, fundamental outlooks.

At first glance, one might think that constancy in the proportions of the three response categories would mean constancy of emissions. But La Branche explains that this is not true. After all, overall emissions reduction, and likely even NET implementation, will become much more commonplace as this century proceeds, simply because of technological advances, public health pressures, regulations, and economic pressures like renewable energy that's cheaper than fossil-fuel energy (as it often is today already). So, later this century, the same three fundamental categories will exist in roughly the same proportions, but in a lower-emissions society. In scientific terms, the distribution curve retains the same shape, but is shifted as a whole to an average lower-emissions level. In consequence, even the most entrenched resisters will emit less than today. The study offers a transport-based example: resisters will refuse to abandon their personal vehicles but later this century will be purchasing more advanced and thus more fuel-economical or even zero-carbon vehicles. The motors in society will have abandoned personal vehicles and will instead be using other types of transport. The variables will be choosing transport modes according to the needs of their life phase. Overall, transport emissions will have dropped sharply—that is, of course, assuming that we ignore the impacts of population growth.

However, reality now rears its ugly head; we have seen that emissions reduction alone will not be enough, no matter how drastic. In fact, we are a good few decades beyond the time when emissions reduction alone might have sufficed. In the current state of affairs, we also need avoidance of new emissions, especially if we do consider population growth, as well as widespread NET implementation. But NETs are not yet available at a meaningful level, so we cannot simply shift the distribution curve for them. Instead, a distribution curve must be generated in the first place. This hints at a need for a genuine step-change transformation of societal attitudes to energy, transport, and general greenhouse gas emissions, rather than shifting the distribution around through odd mixes and matches of partial solutions.

Humanity likely will only converge onto a comprehensive solution once society as a whole, or at least a large majority, comes to realize that Earth's atmosphere-ocean-biosphere veneer is a finite system. We're not at that stage yet. Public opinion polls show that people do care about climate and want to see some action, but also that there is a distinct lack of any sense of urgency, and voting behavior remains dominated by many issues of more immediate concern (Dryzek et al., 2011). As a result, we find ourselves in a period of cunning gamesmanship with all manner of pseudo-green credentials for short-term political or economic gain, rather than real decisive action against climate change. Such gamesmanship includes the potential use of negative emissions as an excuse to delay fossil fuel emissions reduction; the so-called offset game.

It is worthwhile to reflect a little on offsetting. Many offset approaches aim to clean up continued emissions that should instead be stopped, or past emissions. An often-considered example is the following: you (pay someone to) replant a tree, or 1 trillion trees, with the aim of using reforestation as an offset measure. In reality, however, this would at best remove the carbon that was emitted by the original land clearing. So, the true offset capacity is limited, or zero, and it can even fall short. One might argue that there is merit if the original land clearing had happened a long time ago, before we knew any better, and we're now restoring the ecosystem. But offsetting through re-foresting doesn't work when land is used that was cleared only recently, and especially if wild forest was cleared and monoculture forest is to be replanted.

Recently, it was suggested that nature-based solutions are key to achieving Europe's ambitious climate change targets (Bas, 2018). This could lead us down a dangerous path, given that it proposes to use nature-based carbon drawdown mechanisms to offset emissions. Instead, these emissions should be reduced wherever possible, and the nature-based carbon drawdown mechanisms should be employed as important players in a parallel effort to reduce the carbon stock in the climate system, and thus $CO_2$ levels. We cannot double-count the capacity of the carbon drawdown methods. If we deplete that capacity to offset emissions, then we can no longer employ it to reduce $CO_2$ levels.

In short, the key message is that offsetting—through whatever NET—should be considered only for emissions that cannot be avoided or elimin-ated in any other way. This calls for absolutely clarity about how much will be achieved by emissions reduction and how much by carbon removal, in order to avoid a vicious circle in which emissions reduction is postponed for longer and longer, as promises (that may never be achieved at scale) of carbon re-moval grow bigger and bigger (McLaren, 2020).

\* \* \*

Now let's get back to the gist of this section. National governments have been slow to define firm roadmaps for reaching zero, let alone net negative, emissions within decades. In 2015, the best had only promised reductions of 30% by 2030, relative to 2012 emissions levels (*The Guardian*, 2015). More recently, a law-proposal draft has been circulated for an EU-wide 2030 emis-sions reduction target of 50–55% relative to 1990 levels, which unfortunately still won't suffice to get the EU to its net-zero emissions goal by 2050 (Rankin, 2020). Although new initiatives such as large-scale renewable energy imple-mentation, bans on combustion-engine vehicles, and phasing-out of coal power are gradually gaining traction, the timescales for such changes still span many years to decades. Instead of making a step-change in behavior to well-founded new traits, we thus remain trapped in the process of gradually shifting it.

The pace of change is this slow because the environment is not a core business for governments in the same way as the economy, or war, or—as we have seen very recently—a pandemic-causing coronavirus. Such factors have demonstrated capacity to bring swift catastrophe upon government, either through fiscal crisis and punishment by voters or through loss of sovereignty, whereas environmental change has traditionally not had that potential (Dryzek et al., 2011). As a result, vast sums of money were made available in response to the 2008–2009 global financial crisis, whereas nothing like such urgency has ever arisen around environmental issues. If we dare to extrapolate: bringing climate change to the same level of urgency may require unprecedented climatic disasters at apocalyptic scales.

It's frightening to imagine what such disasters would need to look like, given that we have already experienced deadly record heatwaves in Russia and Europe; exceptional drought in the Middle East, parts of Africa, and Australia; extreme flooding in Europe, central and southern Asia, Mozambique and central Africa from Congo to South Sudan; abnormally severe and frequent hurricanes and typhoons; unprecedented intensities and frequencies of mid-latitude storms; marine heatwaves that have become so intense and frequent that they are killing vast stretches of coral-reef habitat; stupendous bushfire emergencies along the US west coast and in Australia and the Amazon; catastrophic glacier-lake collapses wiping out entire valleys after mountain glaciers receded and left massive meltwater lakes behind weak dams of loose rubble moraine; and so on. Astonishingly, all this has failed to bring climate change to the attention of governments with the same urgency as a temporary downturn of the stock markets. Why? Because the dependence of developed countries on continued economic growth to operate and finance a welfare state makes them infinitely more sensitive to financial fluctuations.

To effect a change away from today's focus on selfish human convenience and toward a new focus on working with nature to achieve mutual benefits, serious value needs to be placed on preserving biodiversity and natural habitats, alongside consideration of humanity's ever-expanding footprint on Earth. The importance of ecosystems to society is captured under the term natural capital, comprising:

natural resources from which ecosystems goods and services can be derived. For example, it provides sources of energy, food, and materials;
sinks for wastes and pollution, services of climate, water, and soil regulation; and the environment for living and leisure. In essence, it provides
the core fabric of our societies. ... full accounting for natural capital en-
ables societies to record the full price of our way of life, reveal concealed
debts that are being forwarded to future generations, ... highlight new
ways for economic development and jobs in a green economy based on
green infrastructure, and reframe the base for fiscal revenues and their
use. (European Environment Agency, 2010)

Preserving biodiversity and natural habitats can no longer be seen as a luxury
item that is considered at the tail end of national budgets. Once natural capital is included, GDP no longer is the only measure of economic growth.

If anyone needs convincing that looking after natural capital is of para-
mount importance, then just make them think of it from a selfish point of
view: it is absolutely critical to ensuring future food security for our expanding population. For example, consider the impacts of a collapse of pol-
linator species; of out-of-control plagues of pests and vermin in the case
of natural predator removal; or of collapsed fisheries because of pollution,
ocean current change, overfishing, predator or prey removal, or a global loss
of essential nursery habitats—such as coral reefs—due to warming and acidification along with eutrophication and physical disturbance.

Given that we need all systems to keep functioning and producing if we
are to have a habitable planet, and feed the rapidly expanding population,
we'll need to drastically reduce pollution, curb global warming, and produce
food in ways that sustain the systems, rather than deplete and destroy them.
A powerful illustration was given by Messerli et al. (2019):

Billions of hectares of land have already been degraded, and an additional
12 million hectares of agricultural land are likely to become unusable for
food production every year. Furthermore, agricultural practices can lead
to eutrophication of the aquatic environment, groundwater contamination, soil acidification and atmospheric pollution. Those practices were
also responsible for 60% of the global emissions of $N_2O$ in 2011 ... When
all emissions associated with the global food system are considered,
they account for ... 19 to 29% of total greenhouse gas emissions. Without

technological improvements or other forms of mitigation, especially the restoration of soil health in order to increase its carbon content, greenhouse gas emissions from global agriculture could rise by as much as 87% if production is simply increased to meet the demands of the global population in 2050. That scenario is incompatible with the Paris Agreement and the Sustainable Development Agenda.

The need to protect and grow natural capital cannot be considered in isolation from climate change because processes such as ocean acidification, warming, and increases in extreme weather have massive impacts on habitats and ecosystems. At best, such processes impose severe stress on ecosystems, which makes them much more sensitive to other disturbances. For instance, large-scale, climate-driven forest destruction reduces recovery potential and makes surviving pockets more sensitive to logging and other human activity. At worst, the climatic stressors reach such proportions that they finish entire ecosystems by themselves. A key example concerns ocean acidification and ocean heatwaves, which are on course to destroy coral reefs worldwide (e.g., American Geophysical Union, 2020).

It stands to reason that a drive to maintain healthy habitats and ecosystems should be combined with a drive to reduce wanton exploitation and pollution. This will benefit both public health and wellbeing as well as environmental health. Notably, it will increase natural systems' resilience by rebuilding and sustaining genetic diversity, which in turn will reduce the potential for uncontrollable pestilence outbreaks. Moreover, reduced exploitation and pollution, and improved public health and wellbeing, are essential for a well-functioning, sustainable, and equitable society in which 10 billion humans can coexist.

But it won't come easy. Recent work calculated that an 80% emissions reduction in transportation by 2050 would require a drastic "shift towards light electric vehicles, shift of road freight to electric train, ambitious mineral recycling levels, drastic reductions in the demand for transportation, and a significant decrease in overall economic activity." This in turn requires a "broader socio-economic framework in which current growth-oriented economies evolve towards a new system that fulfils human needs without the necessity for continuous growth" (de Blas et al., 2020). Some governments, most prominently that of New Zealand, have begun to de-emphasize the traditional doctrine of obligatory economic growth in favor of a focus on health and wellbeing (Graham-McLay, 2019; Peat, 2019). However, this concept does not find traction everywhere. A suggestion to consider a more wellbeing-focused budget in Australia was received in Parliament with a sad show of multicultural mockery by the Treasurer (Acharya, 2020). Perhaps casting reform under a term like natural capital would have been more palatable than under wellbeing, but an open mind and some respect would go a long way too.

Actions to deal with climate change need to be carefully evaluated against the need to prioritize reductions in exploitation and pollution. This is important because the rapidly expanding energy transition and transport electrification (Table 7.1), as well as NET and SRM methods, are not a priori environmentally friendly throughout their entire process chain, and especially at the scales that would be required. All require new industrial developments and/or landscape management practices, as well as development of new resource sectors and decommissioning of outdated industries, each of which brings its own environmental and economic costs.

Each of these developments needs to be both economically feasible and acceptable to society and its future priorities if they are to succeed. This calls for attention to detail in identifying and avoiding the possibility

**Table 7.1** Anticipated increase in renewable energy systems from 2015 to 2060.

|  | Stock in society | |
|---|---|---|
|  | 2015 | 2060 |
| Personal vehicles: battery (million) | 1.5 | 1300 |
| Personal vehicles: hybrid, plug-in hybrid, and fuel cells (million) | 14 | 710 |
| Electric bikes (million) | 460 | 1600 |
| Buses: battery electric, hybrid, plug-in hybrid, and fuel cell (million) | 0 | 31 |
| Marine and heavy freight transport: battery electric, hybrid, plug in hybrid, and fuel cell (million) | 0 | 130 |
| Wind power (gigawatts) | 430 | 4200 |
| Solar photovoltaic (gigawatts) | 220 | 6700 |
| Solar thermal (gigawatts) | 21 | 1300 |

*Source*: Månberger and Stenqvist (2018).

that new environmental and public health challenges are created, whether it be immediately or at some stage in the future. Consider, for example, stratospheric $SO_2$ aerosol injections that may exacerbate ozone depletion and acid rain. Or widespread use of toxic elements in electrification that, absent strict recycling procedures, might lead to catastrophic future environmental contamination. Or the use of cobalt in battery technology, which is mined mainly in the Democratic Republic of Congo, Africa, at severe humanitarian and environmental costs (Sanderson, 2019). Under a 100% renewable energy scenario, demand for metals for electric vehicles and renewable energy technologies might even exceed reserves for cobalt, lithium, and nickel (Dominish and Florin, 2019; de Blas et al., 2020). A great summary analysis of this potential for a wide range of materials and technologies, and for a variety of recycling intensities, may be found in Månberger and Stenqvist (2018).

Clearly, it is essential that optimal recycling is achieved on a global scale, both to limit damages and to avoid exceeding supplies. Also, continuing research and development with a focus on material substitution will be vitally important, designing new technologies in which critical materials with limited supplies are replaced by more common materials. The big question that then arises is: who should pay for this wide-ranging package of emissions reduction and avoidance, negative emissions, and reform of society to ensure that these changes occur in a sustainable synergy? To address that, we need to consider the three major players in this game, namely governments, corporations, and consumers.

## 7.2. Governments, Corporations, and Consumers

Markets consist of producers and consumers, and both may change their behavior in ways that reduce emissions. Producers, including large corporations, would move in a low-carbon direction if this optimized their chances to profit from such a shift in society. But there is not much evidence that many corporations will assume a leadership role if governments fail to do so. Even the (re-)insurance sector, which arguably has most to lose from climate change, has produced preciously little comprehensive action (Dryzek et al., 2011). In the cold light of day, corporations will respond when regulations are imposed by government and/or when consumer behavior shows a clear shift that affects the financial bottom line. For governments, and also some corporations, the perceived costs of abating emissions have long been a key cause of sluggish responses. The benefits are often downplayed because most countries or corporations will enjoy only a small share of the total benefits when dealing with a global issue and because the benefits are less directly visible in the budget.

Consumers have become increasingly pro-active, with (geographically variable) uptake of roof-top solar PV, home insulation, growing transport electrification, great increases in public transport use, and so on. Despite those developments, however, a deeply ingrained hankering persists for convenience, comfort, and short-term benefits. This fuels a consumerism that in turn drives an ever-increasing energy dependence per capita, and an ever-increasing total energy need associated with both the per-capita use and the general population growth (Figure 2.5).

Changing government, corporate, and consumer attitudes will require a novel kind of cost:benefit accounting, which assigns serious weight not only to money but also to natural capital, health, and wellbeing. Examples of the wider benefits are: reduction of health hazards from particulate air pollution that is released along with $CO_2$ during fossil fuel combustion; lower costs of living through improved energy efficiency; a higher degree of self-sufficiency as nations rely less on imports of fossil fuels; and a reduction in environmental pollution from operations and accidents associated with fossil-fuel mining and transporting.

A strong case for the sensibleness of shifting to a low carbon-intensity society emerges from a comprehensive analysis of cities (Gouldson et al., 2018). Apart from low-carbon investment in cities returning a potential net value of $16.6 trillion over the period to 2050 (Sudmant et al., 2016), the analysis identifies a wide array of so-called co-benefits, or positive social, economic, and environmental impacts beyond emissions reduction. For example, the report shows that pursuing development of more energy-efficient buildings has positive impacts on public health, employment, and social inclusivity; a push toward low-carbon transport has positive impacts on public health, reduction of congestion and travel times, employment, and social inclusivity; and improved solid-waste management has positive impacts on public health, employment, and social inclusivity. The report argues that:

the benefits of these low-carbon measures extend far beyond emission reductions. The wider economic, social, and environmental impacts may be much more valuable than the financial returns associated with climate action. This bundle of measures could therefore provide a platform for more transformative change by building public enthusiasm for low-carbon urban development, as well as the institutional capacities, financing arrangements, and learning needed for more ambitious action.

This brings us to the concept of the so-called social cost of carbon, which relates to the incremental damage resulting from emitting a ton of carbon dioxide and other greenhouse gases into the atmosphere. It includes factors such as losses and gains to agriculture caused by global warming, flooding from sea-level rise, destruction from more severe tropical cyclones and additional wildfires, health impacts of particulate air pollution, and so on. Hamilton (2017) and Hamilton et al. (2017) report that the lower bound of the average avoided health damages from particles less than 2.5 microns in size (PM2.5) per ton $CO_2$ abated is comfortably above \$36 in all but two (Brazil and South Africa) of the 15 largest emitting nations. Gillingham (2019) reviews literature that extends the social cost beyond just public health and comes to a value of \$50 to \$75 per ton of $CO_2$.

For clarity, these numbers give the amount of financial benefit per 1 ton of $CO_2$ emissions reduction, or 0.27 tons of carbon. In other words, continued or new emission of each ton of $CO_2$ actually costs society this much. The number could therefore be used as a perfectly reasonable template for a tax to be imposed per ton of $CO_2$ emitted. Such a tax would ensure that emissions actually pay for the damage they do to society. By not paying for these damages, fossil fuels are effectively very substantially subsidized. We have touched upon that before, in section 2.5, defining fossil fuel subsidies in terms of "the gap between existing and efficient prices, where efficient prices are those warranted by supply costs, environmental costs, and revenue considerations." The social cost of carbon is a key component of so-called efficient prices.

Let's put some numbers on it, to get a sense of the efficient unsubsidized cost of oil and coal. One barrel of crude oil contains about 159 liters, and each liter today (March 19, 2020) costs about \$0.17. Combustion of a barrel of crude yields 0.43 tons of $CO_2$. The existing, commercial, cost per ton of $CO_2$ emission therefore is just under \$63 per barrel. Coal today goes for \$70 per ton for low grades, to $150 per ton for high grades. Emissions range between about 1 ton of $CO_2$ per ton for low-grade coal and 2.6 tons of $CO_2$ per ton for high-grade coal. Thus, the existing, commercial, cost per ton of $CO_2$ emission from coal is between $57 and $70; very similar to that of a barrel of crude oil. Accounting for the social cost, the efficient costs for both are $50 to $75 higher than the existing, commercial, costs. In effect, therefore, a barrel of crude and a ton of coal should cost roughly double what they cost today, if they were to enter the market as unsubsidized resources that pay for the damages they cause to the environment and public health. In other words, they are currently running on subsidies of about 50%. Who wouldn't want to run a business on such terms?

As long as fossil fuel remains heavily subsidized, many worthy emission abatement projects that can provide both global and local benefits will not be implemented. Hamilton (2017) summarizes the implications as follows: "Reforming public investment management can ensure that social benefits are taken into account and, more generally, that the quality of public investments will increase. In contrast, the private sector has little incentive to reduce production externalities, whether global ($CO_2$) or local (PM2.5). It therefore falls to governments to provide incentives, ideally in the form of taxes on both local and global pollutants, in order to increase the welfare of citizens." In other words, fossil fuels ought to be taxed to a level where they would be effectively unsubsidized and pay themselves for the damages they cause. This would create a level playing field relative to renewable energy sources, which in turn will cause strong market forces toward a transition to renewables, which often are cheaper already. Where fossil fuels remain irreplaceable, market forces would drive rapid increases in energy efficiency.

Gillingham (2019), from another perspective, suggests that "inexpensive measures can be implemented today, including energy conservation, efficiency nudges, and the replacement of retiring fossil-fuel powered electricity generation with renewables. The costs of [implementing] these measures are already lower than the damage from climate change they would avert, based on estimates of carbon's social cost." Either way, inclusion of the social cost of carbon makes it abundantly clear that the time for transition is now and that it makes perfect economic sense with strong returns that (more than) compensate for the investments needed for the transition. Job losses incurred from a transition away from traditional resources will be compensated by new employment in renewables manufacturing, installation, and maintenance.

* * *

Especially when paired with government-imposed regulations—but also in isolation—customer behavior is a critical determinant of the behavior of major corporations by affecting the financial bottom line. Consumer pressure may thus drive corporations to place greater weight on natural capital, health, and wellbeing. This will push them to reduce pollution and excessive exploitation and adopt novel practices to reduce environmental and public-health impacts in general. Major corporations are responding to customer expectations and demands already by rebranding themselves, for example, from British Petroleum to Beyond Petroleum and from petroleum companies to energy companies, or by increasing the manufacturing of electric vehicles and even talking about phasing out vehicles with combustion engines.

Consumers can make direct contributions, some of which can directly influence emissions by producers and providers. Key contributions by consumers are flying less; switching to renewable electricity; installing solar PV and/or a heat pump if they can afford it; driving a smaller car, motorbike, or scooter to minimize fuel requirements and congestion; driving less by instead using public transport, cycling, or walking; insulating homes better; avoiding over-heating and over-cooling by installing better controls and dressing in accordance with the season; eating less meat (especially the resource-intensive beef and lamb; Carrington, 2018b); turning off lights and changing to low-energy lighting; and switching off appliances where possible, rather than leaving them on stand-by. Many of these measures put pressure on producers and providers to change their business models. But consumers can also influence emissions by producers by choosing to reduce consumerism. Even a sustained 10% reduction in replacement of items by consumers—choosing instead avenues to reduce, repair, reuse, or recycle—would create important market pressure toward reduced production. Sustained over decades this would drive a reduction of the throw-away behavior of society and adaptation of corporations toward the manufacturing of more durable and upgradable products. Smartphones provide a good example, with sales—and thus manufacturing and associated emissions—dropping because consumers choose to keep older models for longer—although this comes with a caveat that this trend was driven primarily by consumers' financial considerations rather than environmental concerns (Conwell, 2018). As ever, there is a downside as well; psychologically, recycling is known to create a false sense that more consumption is allowed (Catlin and Wang, 2013). In the end, achieving a big difference to rampant overproduction will require fees or taxes on items and/or imposed regulation on production of items (Wilkins, 2018).

The public in general also sets up major hurdles with respect to the recasting of society and industry into a low carbon-intensity model. Attitudes, beliefs, values, and risk perceptions are critical to societal acceptance of new techniques and industries (Swim et al., 2009). For instance, are people prepared to accept seeing many massive DAC systems dotted around the landscape, given that much less intrusive solar and wind farms often create outrage already? Will the spreading of black and green olivine rock on beaches be palatable? And how about inundating valleys for the installation of more (pumped) hydropower facilities, in direct conflict with an opposite trend to remove existing dams to restore habitat, river ecosystems, and fisheries spawning grounds, and to improve water quality (Cho, 2011)? What to think about the massive tree plantations needed to fuel BECCS facilities, or as a NET in their own right; how could these be implemented without meeting strong opposition both from the food-production sector, from environmental groups concerned with biodiversity and conservation of wild habitats, and from the public on aesthetic grounds? Will farmers and the public accept soil-regeneration practices and possibly a change in the types of food produced and/or their seasonal availability? The list goes on and on.

It is evident that creation of broad public support through information, education, consultation, involvement, and empowerment will be just as important as technological advance. It should include clearly presented results from comprehensive, unbiased environmental impact assessments that emphasize critical issues and dispel myths and inconsistencies. The chances of social acceptability will be greatly enhanced if solutions explicitly address wellbeing and environmental injustice.

Finally, to avoid increasing friction and conflict, decision-making processes will need to be well-informed, respectful, inclusive, and equitable between population groups and nations. As we have seen, poorly informed and unilaterally imposed knee-jerk climate intervention may lead to droughts or floods, famines, conflicts, mass migrations, and disease outbreaks.

Sensitivities to inter-cultural and international inequalities run deep, which is understandable because climate change has been largely caused by industrializing groups and nations that became enriched in the process, whereas poorer groups and nations disproportionally endure the impacts. To avoid repetition of similar injustices, three key requirements emerge. First, a sound foundation of impartial research on climate-intervention methods is needed, to ensure that decisions can and will draw on a basis of best-available evidence and that adverse impacts will be minimized (Meadowcroft, 2009). Second, detailed climate-intervention proposals need to be opened to international assessment and discussion with respect to their cross-border implications and acceptability. Third, a broad and inclusive international framework is needed for identifying climate-intervention methods that might allow one nation to inflict unwanted impacts on other nations, and for developing international governance structures to manage such issues.

## 7.3. Discontent and Litigation

It is a sobering thought that the world seems to be waiting for environmental disasters to become sufficiently large and frequent that they impede economic growth, global finance, and/or the welfare state, before concerted emergency-level action is undertaken to deal with climate change. Maybe no single environmental disaster will ever be big enough to achieve sufficient impact on the global system over a sufficiently long time, although the coronavirus pandemic of 2020 indicates that such impacts are possible. Alternatively, an ever-increasing accumulation of climate disasters may still tip the balance and lead to a decisive call for action. This would likely depend on continued clamoring and civil disobedience by those who do see the writing on the wall, and which gains a bit in intensity with every major event.

Today, only relatively small subsets of the global public consistently express major concerns about climate change. Although they are often visible and vocal, it remains to be seen how effective they can be (Dryzek et al., 2011). Mobilization of inherently large subsets of the public may increase the chances, such as the rallying of global youths around iconic spokespersons like Greta Thunberg (Thunberg, 2019). Although even this offers no guarantee of success against entrenched politics filled with powerful resister-type people, these movements and their impacts also grow steadily with every climate event.

For the movement to really register with the public at large, it needs to overcome the limitations of society's short average attention span. For example, during the 2019–2020 Australian bushfire calamity, which burned 22% of the country's total temperate and broadleaved forest (Boer et al., 2020), virtually the entire world was buzzing about climate change for a good two months. But then the coronavirus burst onto the world stage and the bushfire disaster was all but forgotten on a global scale. Still, let's be positive: its longer-term net legacy appears to have been a boost in numbers of people concerned about climate change and in the intensity of their demands for urgent action against climate change. Even previously denialist Australian government and White House officials now talk about emissions reduction and even NETs in a more nuanced manner than before.

Once public concern and indignation reach a flashpoint, change may occur in several ways. Foremost is through use of the vote in democratic nations, although this leaves a considerable risk that things get bogged down in the trappings of the political process once a new party (or parties) takes over. Political checks and balances are important, but invariably take a lot of time during which preciously little is changed, and we cannot afford to lose that time. In consequence, there are calls to not only focus on the vote but to also ramp up non-violent civil disobedience (McMahon, 2020). Finally, the justice system can be utilized to try and change regulations or ensure that existing regulations are properly enforced (e.g., Ganguly et al., 2018; Bachelet, 2019; Hansen, 2019; Urgenda, 2019; Carrington, 2020). In total, more than 1300 legal actions concerning climate change have been brought since 1990, spread over 28 countries (Laville, 2019).

Climate change related questions of justice range from international to national levels and from intergenerational to personal or company levels. A case might even be made that injustice is being done to nature. International to national levels especially concern "the procedures around which decisions are made, the unfairness of the distribution of existing vulnerabilities to climate change, the fair distribution of benefits and burdens in the present and near future, and the extent and nature of obligations to those within and outside a country" (Dryzek et al., 2011). Intergenerational issues are concerned with each generation's responsibility to future generations, often in terms of what sort of world we should leave our children and grandchildren. At the personal or company level, the focus lies more on direct impacts or damages of climate change to the person or company itself, such as flood or fire damage or crop failures.

Because it is very difficult to relate specific events directly and unambiguously to climate change, there is a high chance of failure with climate lawsuits, especially against governments. This was the case when Electro Optic Systems sued the state of New South Wales following the destruction of the Mount Stromlo Observatory during the 2003 Canberra bushfires; the case was rejected because lightning was the ultimate trigger of the fires (Newhouse, 2020). However, the Urgenda Foundation case in the Netherlands demonstrates that governments can be held to account through the courts. As summarized by Newhouse (2020):

> The Urgenda Foundation fought the Dutch government to force it to reduce the nation's greenhouse gas emissions by 25% from 1990 levels by the end of 2020. In December 2019, the Dutch Supreme Court ruled ... that the government is obligated by the European Convention on the Protection of Human Rights to take suitable measures if a real and immediate risk to people's lives or welfare exists and the state is aware of that risk. And that the obligation to take suitable measures also applies when it comes to environmental hazards that threaten large groups or the population as a whole, even if the hazards will only materialize over the long term. The case marked the first time a government has been required by the courts to take action against climate change.

The Urgenda case, initiated in 2013 and finally won in 2019, focused on general impacts from a lack of following certain agreements, rather than on damage done by a specific event. Similar legal strategies are being pursued in Canada, France, Germany, India, New Zealand, the United Kingdom, and the United States. At around the same time as the initial verdict in the Urgenda case, the case of Ashgar Leghari vs. The Federation of Pakistan was concluded, with the Lahore High Court ordering the Pakistani government to appoint a focal person on climate change and develop a list of adaptation measures to be implemented by 2015 (Ganguly et al., 2018). Personally, I have had some involvement in the compilations of scientific support to two cases in the United States (Hansen et al., 2013, 2017). The first was Alec L., et al. vs. G. McCarthy et al., which was filed in May 2011 and finally denied a hearing by the US Supreme Court in December 2014, and the other is Juliana vs. United States, filed in October 2014 and still wrangling its ways through the courts (Our Children's Trust, 2020a,b). These various cases illustrate that the time involved in seeking action by governments though the judiciary easily ranges across 3–6 years or more from initiation to final outcome. Next, following a positive verdict, many more years will pass before meaningful changes to total emissions will be observable.

Apart from suing governments for failing to take steps to mitigate climate change, three other key types of litigation are emerging (Korbel, 2019). The first type seeks to challenge decisions that approve projects and developments, such as coal mines, coal-fired power stations, and oil and gas exploration. The second type seeks to force companies directly and indirectly affected by climate change risks to completely and transparently evaluate and report on those risks. The objectives then are to protect investors in those businesses from either direct physical climate risks or financial losses associated with transition to a low-carbon economy, or to litigate against companies responsible for significant emissions. The third type concerns litigation against companies responsible for significant direct or indirect greenhouse gas emissions, with an aim of bringing injunctions to prevent further emissions. A new category of cases within the third type concerns misleading conduct by major emitters, when they are found to have previously recognized the risks of climate change and their contribution to it, but then actively sought to undermine public recognition of it. The 1980s Exxon and Shell reports on climate change in response to $CO_2$ emissions come to mind here (section 2.5) (see also Irfan, 2019).

The landscape of litigation will undoubtedly develop rapidly, including many yet unforeseen directions. But one thing is sure because of all the due diligence, back-and-forth challenges and appeals, questions of jurisdiction, time taken to follow up on decisions, and so on: it's a slow process (Irfan, 2019). Thus, it will take many years or even decades for this approach to take a significant bite out of the problem. And this is time we can ill afford, given that current emissions are adding about 10 GtC per year to the climate system, and emissions and atmospheric $CO_2$ concentrations are still increasing every year, rather than stabilizing or decreasing (Friedlingstein et al., 2019; Johnson, 2019; $CO_2$-Earth, 2020).

Given the increasing urgency of actions against climate change, it is interesting to note that one specific litigation strategy is thought to have a so-called magic-bullet potential. It is the type that seeks direct civil liability against those responsible for major greenhouse gas emissions. Its magic-bullet potential comes from the fact that this type of litigation can offer a means of regulation by itself, when any finding of liability triggers a cascade of major emitters scrambling to avoid the unwelcome spotlight (Ganguly et al., 2018; and references therein). Those authors furthermore concluded that new private climate change lawsuits may be more successful because of the rapidly evolving scientific, discursive, and constitutional context, which generates new opportunities for judges to rethink the accountability of major private carbon producers. Meanwhile, even unsuccessful cases help articulate climate change as a legal and financial risk. Overall, a clear message emerges that there is substantial merit in pushing hard through the judicial route.

## Key Sources and Further Reading

Acharya, M. Josh Frydenberg criticised for "racially abusive" speech with references to symbolic Hindu practices. *SBS*, February 28, 2020. https://www.sbs.com.au/language/english/audio/josh-frydenberg-criticised-for-racially-abusive-speech-with-references-to-symbolic-hindu-practices

American Geophysical Union. Warming, acidic oceans may nearly eliminate coral reef habitats by 2100. *ScienceDaily*, February 18, 2020. https://www.sciencedaily.com/releases/2020/02/200218124358.htm

Bachelet, M. Bachelet welcomes top court's landmark decision to protect human rights from climate change. United Nations Human Rights, Office of the High Commissioner, December 20, 2019. https://www.ohchr.org/EN/NewsEvents/Pages/DisplayNews.aspx?NewsID=25450&LangID=E

Baker, D. 2020. As in Chapter 4.

Bas, L. Nature-based solutions are key to achieving Europe's ambitious climate change targets. IUCN, October 8, 2018. https://www.iucn.org/news/europe/201810/nature-based-solutions-are-key-achieving-europes-ambitious-climate-change-targets

Boer, M.M. Unprecedented burn area of Australian mega forest fires. *Nature Climate Change*, February 24, 2020. https://www.nature.com/articles/s41558-020-0716-1

Carrington, D. Can climate litigation save the world? *The Guardian*, March 20, 2018a. https://www.theguardian.com/environment/2018/mar/20/can-climate-litigation-save-the-world

Carrington, D. Avoiding meat and dairy is "single biggest way" to reduce your impact on Earth. *The Guardian*, June 1, 2018b. https://www.theguardian.com/environment/2018/may/31/avoiding-meat-and-dairy-is- single-biggest-way-to-reduce-your-impact-on-earth

Carrington, D. Heathrow third runway ruled illegal over climate change. *The Guardian*, February 28, 2020. https://www.theguardian.com/environment/2020/feb/27/heathrow-third-runway-ruled-illegal-over-climate-change

Carroll, R. 2019. As in Chapter 4.

Catlin, J.R., and Wang, Y. Recycling gone bad: when the option to recycle increases resource consumption. *Journal of Consumer Psychology*, 23, 122–127, 2013. https://www.sciencedirect.com/science/article/abs/pii/S1057740812000381

Cho, R. *Removing dams and restoring rivers*. State of the Planet—Earth Institute, Columbia University, August 29, 2011. https://blogs.ei.columbia.edu/2011/08/29/removing-dams-and-restoring-rivers/

Conwell, S. Smartphone sales are slowing and here are two key reasons why. *CNBC*, February 24, 2018. https://www.cnbc.com/2018/02/23/smartphone-sales-are-slowing-and-here-are-two-key-reasons-why.html

$CO_2$-Earth. *$CO_2$ Acceleration*. January 8, 2020. https://www.co2.earth/co2-acceleration

Cox, E.M., et al. Blurred lines: the ethics and policy of greenhouse gas removal at scale. *Frontiers in Environmental Science*, 6, 38, 2018. https://www.frontiersin.org/articles/10.3389/fenvs.2018.00038/full

de Blas, I., et al. The limits of transport decarbonization under the current growth paradigm. *Energy Strategy Reviews*, 32, 100543, 2020. https://www.sciencedirect.com/science/article/pii/S2211467X20300961

Dominish, E., and Florin, N. Electric cars can clean up the mining industry—here's how. *The Conversation*, April 17, 2019. https://theconversation.com/electric-cars-can-clean-up-the-mining-industry-heres-how-115369

Dryzek, J.S., et al. Climate change and society: approaches and responses. In: Dryzek, J.S., et al. (eds.) *The Oxford Handbook of Climate Change and Society*, Oxford University Press, Oxford, United Kingdom, pp. 3–17, 2011. https://www.oxfordhandbooks.com/view/10.1093/oxfordhb/9780199566600.001.0001/oxfordhb-9780199566600-e-1?print=pdf

European Environment Agency. Chapter 8: future environmental priorities: some reflections. In: Martin, J., et al. (eds.) *The European environment—state and outlook 2010: synthesis*. European Environment Agency, Copenhagen, Denmark, pp. 151–164, June 3, 2010. https://www.eea.europa.eu/soer/synthesis/synthesis/chapter8.xhtml

Friedlingstein, P., et al. Global carbon budget 2019. *Earth System Science Data, 11,* 1783–1838, 2019. https://www.earth-syst-sci-data.net/11/1783/2019/essd-11-1783-2019.pdf

Ganguly, G., et al. If at first you don't succeed: suing corporations for climate change. *Oxford Journal of Legal Studies, 38,* 841–868, 2018. https://academic.oup.com/ojls/article/38/4/841/5140101

Gillingham, K. 2019. As in Chapter 3.

Gouldson, A., et al. The economic and social benefits of low-carbon cities: a systematic review of the evidence. Coalition for urban transitions. London and Washington DC, June, 2018. https://ledsgp.org/wp-content/uploads/2018/06/The-Economic-and-Social-Benefits-of-Low-Carbon-Cities-A-systematic-review-of-the-evidence.pdf

Graham-McLay, C. New Zealand's next liberal milestone: a budget guided by "well-being." *The New York Times,* May 22, 2019. https://www.nytimes.com/2019/05/22/world/asia/new-zealand-wellbeing-budget.html

Hamilton, K. Economic co-benefits of reducing $CO_2$ emissions outweigh the cost of mitigation for most big emitters. *Grantham Institute News,* November 3, 2017. http://www.lse.ac.uk/GranthamInstitute/news/economic-co-benefits-of-reducing-co2-emissions-outweigh-the-cost-of-mitigation-for-most-big-emitters/

Hamilton, K., et al. Multiple benefits from climate change mitigation: assessing the evidence. The Grantham Research Institute on Climate Change and the Environment, November 2017. http://www.lse.ac.uk/GranthamInstitute/wp-content/uploads/2017/11/Multiple-benefits-from-climate-action_Hamilton-et-al-1.pdf

Hansen, J. Wheels of justice. December 26, 2019. http://www.columbia.edu/~jeh1/mailings/2019/20191226_WheelsOfJustice.pdf

Hansen, J., et al. 2013. As in Chapter 2.

Hansen, J., et al. 2017. As in Chapter 2.

Irfan, U. Pay attention to the growing wave of climate change lawsuits. *Vox,* June 4, 2019. https://www.vox.com/energy-and-environment/2019/2/22/17140166/climate-change-lawsuit-exxon-juliana-liability-kids

Johnson, S.K. Here's how much global carbon emission increased this year. *Ars Technica,* June 12, 2019. https://arstechnica.com/science/2019/12/2019-carbon-emissions-look-to-tick-upwards-again/

Korbel, A. A new era of climate change litigation in Australia? *Corrs Chambers Westgarth,* April 8, 2019. https://corrs.com.au/insights/a-new-era-of-climate-change-litigation-in-australia

La Branche, S. Embedding the social sciences in a long-term carbon-neutrality scenario: perspectives on a 2050 zero net emissions study. World Resources

Institute, Expert Perspectives. Accessed February 23, 2020. https://www.wri.org/climate/expert-perspective/embedding-social-sciences-long-term-carbon-neutrality-scenario

Laville, S. Governments and firms in 28 countries sued over climate crisis—report. *The Guardian*, July 4, 2019. https://www.theguardian.com/environment/2019/jul/04/governments-and-firms-28-countries-sued-climate-crisis-report

Månberger, A., and Stenqvist, B. Global metal flows in the renewable energy transition: exploring the effects of substitutes, technological mix and development. *Energy Policy, 119*, 226–241, 2018. https://www.sciencedirect.com/science/article/pii/S0301421518302726

McLaren,D.Carbonremoval:thedangersofmitigationdeterrence.*C2G*,March17,2020. https://www.c2g2.net/carbon-removal-the-dangers-of-mitigation-deterrence/

McMahon, J. Former UN climate chief calls for civil disobedience. *Forbes*, February 24, 2020. https://www.forbes.com/sites/jeffmcmahon/2020/02/24/former-un-climate-chief-calls-for-civil-disobedience/#3286d4b13214

Meadowcroft, J. 2009. As in Chapter 6.

Messerli, P., et al. *The future is now: science for achieving sustainable development. Global sustainable development report 2019*, United Nations, Department of Economic and Social Affairs, New York, 245 pp., 2019. https://sustainabledevelopment.un.org/content/documents/24797GSDR_report_2019.pdf

Mommers, J. Lawyers are going to court to stop climate change. And it might just work. *The Correspondent*, December 19, 2019. https://thecorrespondent.com/185/lawyers-are-going-to-court-to-stop-climate-change-and-it-might-just-work/9957255-1fcd8336

Newhouse, G. I've won cases against the government before. Here's why I doubt a climate change class action would succeed. *The Conversation*, January 15, 2020. https://theconversation.com/ive-won-cases-against-the-government-before-heres-why-i-doubt-a-climate-change-class-action-would-succeed-129707

Our Children's Trust. *Federal climate lawsuit filed in 2011 (Alec L. v. McCarthy)*. Accessed March 6, 2020a. https://www.ourchildrenstrust.org/aleclvmccarthy

Our Children's Trust. *Juliana v. United States youth climate lawsuit*. Accessed March 6, 2020b. https://www.ourchildrenstrust.org/juliana-v-us

Peat, J. Economic growth is an unnecessary evil, Jacinda Ardern is right to deprioritise it. *The London Economic*, May 31, 2019. https://www.thelondoneconomic.com/opinion/economic-growth-is-an-unnecessary-evil-jacinda-ardern-is-right-to-deprioritise-it/31/05/

Rankin, J. EU member states call for 2030 climate target. *The Guardian*, March 3, 2020. https://www.theguardian.com/world/2020/mar/03/eu-member-states-call-for-2030-climate-target

Sanderson, H. Congo, child labour and your electric car. *Financial Times*, July 7, 2019. https://www.ft.com/content/c6909812-9ce4-11e9-9c06-a4640c9feebb

Sudmant, A., et al. Low carbon cities: is ambitious action affordable? *Climatic Change*, *138*, 681–688, 2016. https://link.springer.com/article/10.1007/s10584-016-1751-9

Swim, J., et al. Psychology and global climate change: addressing a multi-faceted phenomenon and set of challenges. Report of the American Psychological Association task force on the interface between psychology and global climate change. American Psychological Association, 2009. https://www.apa.org/science/about/publications/climate-change

*The Guardian*. Which countries are doing the most to stop dangerous global warming? October 16, 2015. https://www.theguardian.com/environment/ng-interactive/2015/oct/16/which-countries-are-doing-the-most-to-stop-dangerous-global-warming

Thunberg, G. *No one is too small to make a difference*. Penguin Books, London, 68 pp., 2019. ISBN: 978-0-141-99174-0

Urgenda. *Landmark decision by Dutch supreme court*. December 20, 2019. https://www.urgenda.nl/en/themas/climate-case/

Wilkins, M. More recycling won't solve plastic pollution. *Scientific American*, July 6, 2018. https://blogs.scientificamerican.com/observations/more-recycling-wont-solve-plastic-pollution/

# 8
# The Future
## Toward Rebalancing Climate

We have at last come to the point from which a vision for the future can be discussed, drawing together information on emissions reduction and avoidance, implementation of negative emissions technologies, solar radiation management, adaptation, and the human dimension of how to drive change.

The first, critical question that needs an answer is whether we can still avoid dangerous warming above the Paris Agreement's limits of 1.5°C and 2°C. This question requires quick reminder with respect to the information shown in Figure 2.5. It revealed a pathway of very high emissions in a business-as-usual scenario with no (0%) annual reduction of emissions, driven by both the increase in population numbers and an increasing energy demand per capita. By 2100, this would lead to atmospheric $CO_2$ levels at around 910 ppm and global warming in excess of 5°C above the 2018 temperature. This is a disastrous scenario with catastrophic ocean warming and acidification; ice-free conditions in the Arctic and around Antarctica; a rapid pathway toward virtually complete polar and mountain ice melt and associated sea-level rise that will build up over many centuries toward several tens of meters above the present level; complete disintegration of permafrost with widespread methane release; global extinction of coral reefs; comprehensively disturbed food webs both on land and in the oceans; vast tracts of land where heat stress and heat death conditions dominate over long periods of the year; and so on.

Figure 2.5 also shows that very severe measures of global emissions reduction and avoidance would be needed, decreasing by 4% every year from today, to stay below a 2°C warming limit and that truly draconian global emissions reduction of 7–8% per year would be required to stay below a 1.5°C warming limit. Note that those numbers were based on emissions reduction and avoidance measures only. Also, it was assumed that those measures started from 2018. For every year emissions are not reduced, the feasibility of reaching below 2°C or 1.5°C decreases rapidly, as more than 10 billion tons of carbon are added to the climate system every year.

Although the novel coronavirus impact on the global economy is continuing to develop, and we can't see the full impact yet, it was forecast early on that we may expect it to result in a 5–10% annual reduction in global emissions for 2020. Over the first half of 2020, the observed reduction was 8.8% (Liu et al., 2020). It is a massive blow, associated with the deepest recession on record. Clearly, to be economically sustainable at a global level over many years, such emissions reduction levels would require a rigorous overhaul of how society does things. We looked at ways of doing this in Chapter 3, but it is doubtful that a 5–10% reduction per year can be reached and sustained for decades. Given that in recent years, prior to the coronavirus impact, emissions were still increasing year on year, even a 2% per year decrease of global carbon emissions seems a serious challenge. But it's a challenge that seems to come close to realistic projections for the future.

As we saw in section 3.4, the Bloomberg New Energy Finance (2019) report's projection based on current economic and development trends suggests that emissions will be kept constant at current emissions (about 10 GtC per year) until 2026 and then drop by about 2% per year. Alarmingly, the 2% per year decrease of global carbon emissions comes quite late in that projection, and we calculated in section 3.4 that the Bloomberg projection would cause the emission limit for the 1.5°C warming target to be broken as early as 2039 and that for the 2°C target by about 2075. The 2% per year reduction in global carbon emissions really must be started earlier than 2026. In magnitude, an annual 2% carbon emissions reduction is equivalent to shutting down 80% of all 2018–2019-level air travel. In reality, of course, the airline industry doesn't have to account for the entire 2%. Major gains are possible, and should be made, from transiting to renewable energy and through reducing $CO_2$ emissions from the manufacturing industry. And this brings us to the potential role of the additional approach we have discussed in Chapter 4: NETs.

NETs capture $CO_2$ at source—the industrial smokestack—or at some later stage from the open atmosphere or ocean, and then remove the $CO_2$ by burying it in some form. The resultant negative emissions might be used either to offset ongoing emissions—but remember the issues with offsetting discussed in section 7.1—or to actually reduce the carbon inventory of the climate system. I synthesized key information for them in Figures 4.12 and 4.13. Figure 8.1 refreshes the main points (panels a and b), and then adds

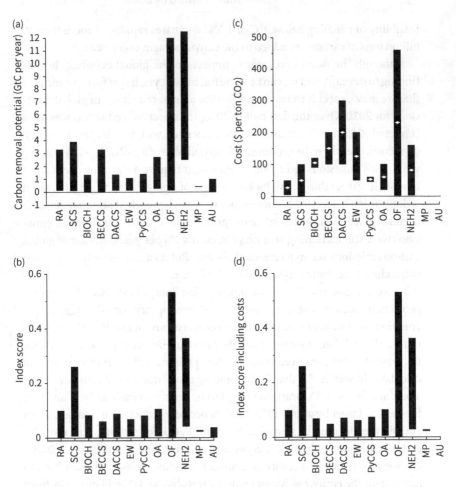

**Figure 8.1** Summary evaluation for NETs including cost information. **a**. Potential capacities for carbon removal, after Figure 4.13. **b**. Index for technical and environmental feasibility, after Figure 4.13. **c**. Summary of current costs after Fuss et al. (2018), Rau et al. (2018), Fuhrman et al. (2019), and Schmidt et al. (2019). PyCCs costs are taken as half those of Biochar because PyCCS provides recovery of more product that may be monetized. By definition, commercial production of mineralized products (MP) must pay for itself, and the costs of artificial upwelling are unknown, so this method is left blank. **d**. Index values after inclusion of the costs, as outlined in Table 8.1.

information about cost estimates of the various NETs in today's money (panel c). Panel d then shows a re-calculated index that includes the cost information (see also Table 8.1). This results in almost imperceptible reorganizations in comparison with the index without costs (panel b); the overall message doesn't change.

In any case, I prefer not to let current cost estimates carry too much weight, because costs are a fickle thing; economies of scale, technological breakthroughs, and shifts in resource availability can rapidly and dramatically change costs through time. In the early days, solar PV and wind power were prohibitively expensive, and cost projections never showed the rapid reduction that was eventually realized. Any technology that seems out of reach today could suddenly become extremely attractive because of—for example—a yet-unforeseen technological breakthrough. So, as long as the

**Table 8.1** Index values and cost estimates of various NETs.

|  | Index Best | Index Worst | Cost Best ($/tCO2) | Cost Worst ($/tCO2) | Cost index Best | Cost index Worst | Total index with cost Best | Total index with cost Worst |
|---|---|---|---|---|---|---|---|---|
| RA | 0.099 | 0.000 | 5 | 50 | 0.99 | 0.90 | 0.098 | 0.000 |
| SCS | 0.260 | 0.007 | 0 | 100 | 1.00 | 0.80 | 0.260 | 0.005 |
| BIOCH | 0.083 | 0.000 | 90 | 120 | 0.82 | 0.76 | 0.068 | 0.000 |
| BECCS | 0.060 | 0.000 | 100 | 200 | 0.80 | 0.60 | 0.048 | 0.000 |
| DACCS | 0.088 | 0.004 | 100 | 300 | 0.80 | 0.40 | 0.070 | 0.002 |
| EW | 0.068 | 0.006 | 50 | 200 | 0.90 | 0.60 | 0.061 | 0.003 |
| PyCCS | 0.081 | 0.000 | 45 | 60 | 0.91 | 0.88 | 0.073 | 0.000 |
| OA | 0.105 | 0.006 | 20 | 100 | 0.96 | 0.80 | 0.101 | 0.005 |
| OF | 0.533 | 0.001 | 2 | 460 | 1.00 | 0.08 | 0.531 | 0.000 |
| NEH2 | 0.365 | 0.042 | 3 | 160 | 0.99 | 0.68 | 0.362 | 0.028 |
| MP | 0.027 | 0.022 | 0 | 1 | 1.00 | 1.00 | 0.027 | 0.022 |
| AU | 0.037 | 0.000 |  |  |  |  |  |  |

*Notes*: Index values as calculated in Table 4.4 (first two index columns), minimum and maximum NETs costs estimates in $ per ton $CO_2$, complement of normalized cost index values (in 1−cost/500, so that 0 is worst and 1 is best), and total index values as plotted in Figure 8.1. Total index values are simple linear products of index values × cost index values. Costs are as reported at the time of publication, without any future indexing, discounting, or reductions due to economies of scale. For the cost sources, see caption of Figure 8.1.

costs are not astronomical, costs per se don't appear to be very important when ranking NETs relative to one another. Another financial aspect, however, will play a critical role: the potential for commercialization, based on the potential to monetize a technology's core deliverable and/or any associated benefits, which are known as co-benefits. Two key questions need answering. First: what do we need to get a technology to realization? Second, what benefit, or utility, will we get from it? It is easier to start answering these questions with the second one, which especially determines the potential for uptake of a new technology.

We have seen that soil-, biochar-, and vegetation-based NETs carry co-benefits that range across increased agricultural yield, improved water quality and availability in the landscape, improved soil stability and thus reduced hazards, reduced need to invest in artificial fertilizers, and so on. Among the NETs of ocean fertilization, alkalinization, and negative-emissions hydrogen NETs, we have encountered potential co-benefits from improved marine biodiversity and conservation, fisheries, reef preservation that helps with coastal protection, and hydrogen production. Other co-benefits we have discussed are trading profits from products made from captured carbon and industrial growth and employment opportunities. In addition to such co-benefits, all NETs offer the core benefit of carbon drawdown. This by itself will become an important financial driver once fiscal systems are established to discourage emissions and encourage drawdown.

Once a technology's rewards from core benefits and co-benefits exceed the costs associated with its drawbacks and initial investments, widespread adoption and implementation can lead to cost implosions similar to those experienced with solar PV and wind power. For most NETs, this potential cannot really be predicted yet because they are at such early stages of their development curves. Similarly, a currently unknown method might be discovered that is then found to have very favorable attributes, and which accordingly starts to dominate NETs development. NEH2, for example, has been proposed only recently. Initial signs for its capacity are promising, but its success may hinge less on its potential capacity than on the financial feasibility for companies to engage in it. It might rapidly become dominant if the world transitions to a hydrogen-based economic model, or at least one in which hydrogen is an important contributor, because great demand for hydrogen would then make NEH2 financially interesting. Conversely, it might fail if the world instead transitions to a direct electricity and bio- or synthetic-fuels model because a continued lack of demand for hydrogen then renders widespread NEH2 implementation unprofitable.

Overall, to bring NETs into the levels of readiness required by humanity's gigatons challenge, intensive programs of research, development, and innovation are needed, followed by a commercialization approach that caters specifically for each individual NET's peculiarities. Targeted government incentive schemes may help with initial investment requirements. To achieve enduring success, solutions need to mature to commercially viable status, where each method's benefits and co-benefits should offset early investments and (re-)mediation of adverse effects. Any future carbon drawdown revenues, or avoided penalties, need to be factored into the benefits. Once economic viability is achieved, initial government incentives can be pared down. Optimization of the pathway to implementation and commercialization requires careful connection of initial research stages of concept development and small-scale proof-of-concept with follow-up efforts toward up-scaling and implementation. For these follow-up efforts, partnerships between the research sector and existing or emerging industries will be essential. Eventually, at-scale implementation will reduce unit costs, and the technology meanwhile expands employment opportunities, skills development, and industrialization.

* * *

Climate mitigation studies typically consider the future development of NETs using so-called Integrated Assessment Models. These are coupled models of the global economic and climate systems that represent fossil fuel emissions from energy generation as well as land use change and forestry emissions and non-$CO_2$ greenhouse gas emissions. At their core, Integrated Assessment Models are economic models adapted and used for making projections about technology deployments, carbon prices, and emissions. They help explore the impacts of social discount rates and rates of technological change, and thus are valuable for assessing the implications and tradeoffs of meeting the aggressive targets needed to limit catastrophic climate damages (Fuhrman et al., 2019). Those authors make three important observations. First, scenarios to limit warming to 1.5°C generally require NETs deployment on the order of 30–50% of current emissions (roughly 3–5 GtC per year). Second, Integrated Assessment Models to date have extensively considered only bioenergy with carbon capture and storage (BECCS) and reforestation and afforestation (RA). Other NETs were omitted because they connect to sectors that are not yet included in the models and/or because too little is known about these technologies in the absence of commercial deployment. A third key observation is that NETs "follow a logistic growth path, where exponential growth gives way to similar slow-down of growth to a constant level of deployment." As such, NETs follow the typical growth pattern of most major technological breakthroughs (Grübler et al., 1999). This offers a useful approximation for making some indicative projections into the future for a wider variety of NETs, which I will do in the following.

To create an indicative assessment of the influence of the entire portfolio of NETs, rather than just two or three of them, I will represent the development and implementation curve for each with a sigmoidal, logistic-style function. The function form used is the same for each NET, except that I define for each NET its individual amplitude of carbon removal based on the mean and full range shown in panel a of both Figure 4.13 and Figure 8.1 (values summarized in Table 8.2) and assign the individual duration of development of each

**Table 8.2** Mean and full range values used in projections with logistic functions.

| NET | TRL | Mean τ | ± τ range | Mean GtC/y | ± GtC/y range |
|---|---|---|---|---|---|
| RA | 9 | 30 | 15 | 1.72 | 1.58 |
| SCS | 9 | 30 | 15 | 2.055 | 1.945 |
| BIOCH | 7 | 50 | 25 | 0.675 | 0.665 |
| BECCS | 5 | 70 | 35 | 1.65 | 1.65 |
| DACCS | 8 | 100 | 50 | 0.745 | 0.605 |
| EW | 6 | 150 | 75 | 0.615 | 0.475 |
| PyCCS | 5 | 70 | 35 | 0.705 | 0.695 |
| OA | 3 | 90 | 45 | 1.485 | 1.215 |
| OF | 3 | 90 | 45 | 6.07 | 5.93 |
| NEH2 | 2 | 100 | 50 | 8.75 | 6.25 |
| MP | 6 | 60 | 30 | 0.24 | 0.2 |
| AU | 2 | 100 | 50 | 0.5 | 0.5 |

*Note*: Timescale τ for each NET is the period over which full impact is reached and is set according to TRL level (Table A2.1 in Appendix 2), except in the cases of DACCS and EW. DACCS timescale, after Fuhrman et al. (2019). EW timescale based on 100 times accelerated natural weathering rate of 0.021% (Lefebvre et al., 2019).

NET to its full impact based on the technological readiness levels (Table A2.1 in Appendix 2), except in the cases of direct air capture and carbon storage (DACCS) and enhanced weathering (EW) (Table 8.2). For details, including the exact equation used, see Appendix 2. In essence, this assessment uses a more rapid development curve to full potential for cheaper, ready-to-go technologies. This is especially true for methods that deliver associated co-benefits such as products of crops that can be independently monetized and thus favorably affect the economic outlook of the method. More expensive, technologically demanding or legally challenging methods—for example, those depending on release of materials into the oceans—have slower development curves to full potential.

My assessment does not include biochar (BIOCH) and BECCS because these would double count carbon drawdown that is covered already by pyrogenic carbon capture and storage (PyCCS) and RA, respectively. PyCCS is a different way of doing the same as BIOCH does, but it is potentially more cost-efficient because it delivers more products that can be monetized. RA may be time-limited in its full potential for carbon drawdown, as it is the establishing of new biomass that provides net carbon drawdown, but BECCS would harvest a large part of the RA biomass and then provide a continuous cycling that effectively keeps the benefit of RA at high/full potential. Hence, my assessment rolls the two methods into one under the RA label.

In the top panel of Figure 8.2, the sigmoidal (S-shaped) growth functions for different NETS are shown in a cumulative manner, based on the mean values for carbon drawdown and development timescale. On land, RA and soil carbon storage (SCS) are among the methods with the largest potential that are immediately available; they have high technological readiness levels. DACCS is a method that may have high potential capacity but is generally reported at more modest levels in realistic estimates, as reflected in Figure 8.1a, and it also needs quite a lot of time to achieve at-scale implementation. Among the marine-based methods, ocean fertilization (OF) and negative emissions hydrogen (NEH2) have very high estimated capacities, which need further constraining through in-depth investigation, but they also have long development timescales.[1]

In the bottom panel of Figure 8.2, one scenario shows the mean and 68% confidence envelope assuming that all NETs considered come to full fruition. A second scenario shows the mean and 68% confidence envelope for all NETs minus OF and NEH2. This second scenario evaluates the implications

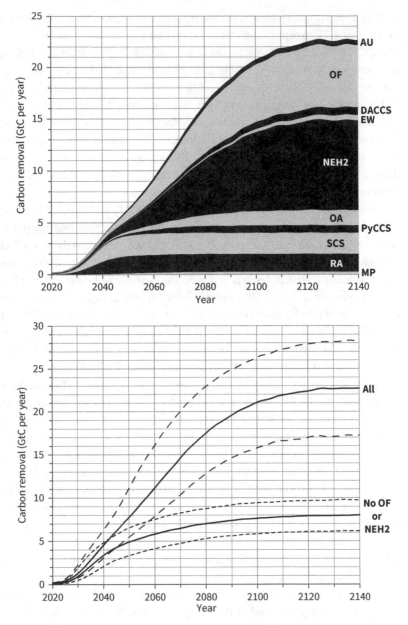

**Figure 8.2** Cumulative carbon removal for the various NETs over the next 120 years.

**Figure 8.2** Continued

**Top:** cumulative carbon removal from the climate system (in GtC per year) based on exponential growth and then slowdown to mean values for the different NETs considered. BIOCH and BECCS are not included because they would double count drawdown covered by PyCCS and RA already. The cumulative values are determined by adding effects of NETs in order of increasing mean cost values, as shown with white diamonds in panel c of Figure 8.1

**Bottom:** cumulative carbon removal from all methods shown in the top panel and from all minus the potentially large marine NETs of OF and NEH2. For both projections, means are shown (solid lines) with their respective 68% confidence envelopes (dashed lines).

For details behind the construction of these diagrams, see Appendix 2.

of large-scale marine NETs not coming to fruition. This is possible because all are in early stages of development, or essentially unknown, and because they may face significant legal hurdles before upscaling could even be considered. I then combine these two potential extreme scenarios for the future of NETs with the emissions reduction scenarios of Figure 2.5, to offer indicative outlooks of atmospheric carbon accumulation and drawdown, and temperature (Figure 8.3).

The projections in Figure 8.3 should not be confused with the much more elaborate and interactively determined projections from Integrated Assessment Models. All my projections do is provide indicative insight into the role and importance of different measures for climate change mitigation. Nevertheless, my projections have one major benefit: they offer a glimpse at how important, or insignificant, different NETs can be. The more complex models still don't account for the full gamut of NETs and thus cannot offer that insight. I show results for the two extreme scenarios in Figure 8.3. One is a more realistic scenario, in which certain important NETs fail to come to fruition, leaving a lesser overall NETs capacity (panel a). The other is a decidedly optimistic scenario, in which all NETs manage to live up to their promise (panel b).

Interestingly, the timescales involved in getting any NETs to full operation imply that both scenarios shown in Figure 8.3 require emissions reduction by 4% per year (solid lines) to give us any chance of keeping warming over the next century below about 1.5°C, at the 68% confidence level. By around

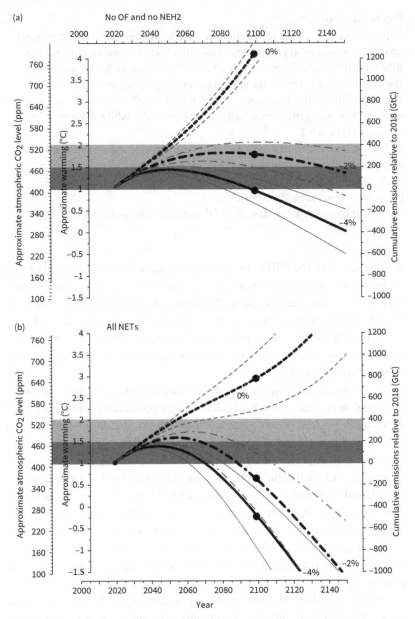

**Figure 8.3** Indicative projections of cumulative carbon emissions and approximate atmospheric $CO_2$ levels and surface temperatures. These projections use the carbon emissions reduction scenarios of 0% (based on

**Figure 8.3** Continued

historical growth), −2%, and −4% from Figure 2.5 in Chapter 2. The projections then account for carbon removal based on the two scenarios shown in the bottom panel of Figure 8.2.

**a.** Solutions using carbon removal by the NETs portrayed in the top panel of Figure 8.2, except for the major marine ones (OF and NEH2). Effectively, this scenario shows the implications of full on-land NETs implementation but largely failed marine NETs.

**b.** Solutions using carbon removal by all NETs portrayed in the top panel of Figure 8.2.

For each solution, the mean is indicated with a heavy line and the 68% confidence bounds are shown with light lines. The dark grey band indicates warming to the Paris Agreement's 1.5°C limit. The grey band extends this to the 2.0°C limit. Heavy dots identify the mean values in 2100.

2100, the scenario in which certain important NETs fail to come to fruition (panel a) may after the initial warming up to 1.5°C return toward temperatures similar to those of today. In the optimistic case with all NETs working to capacity (panel b), temperatures may even drop a degree or slightly more below today's values, although the initial warming would be rather similar to that in the case shown in panel a.

Emissions reduction of 2% per year (dash-dash-dot lines) still manages to keep temperatures at or below 2°C in both scenarios. The scenario in which certain important NETs fail to come to fruition sees temperatures higher than the 1.5°C limit of the Paris Agreement by 2100. The optimistic scenario, however, provides cooling of about 0.5°C relative to the present.

In the absence of emissions reductions (dashed 0% lines), both scenarios exceed 2°C warming by 2050 for the mean values and by or well before 2080 for the 68% confidence envelopes. Moreover, temperatures will continue to increase thereafter. Hence, application of NETs without emissions reduction and avoidance is not a sensible option.

But even if we manage to implement comprehensive portfolios of NETs along with emissions reduction and avoidance, the future is not going to be plain sailing, unfortunately. This is because of slow responses and feedbacks in the climate system, and also because of tipping points.

First, let's look at slow responses and feedbacks. While the strong downward curvature of the lines with 2% or 4% emissions reduction looks promising, it is concerning that initially there is warming up to 1.5°C or even more, until at least 2050. This means that slow response and feedback processes in the climate system will continue to get further and further out of equilibrium, which sets them up for centuries of ongoing response even when temperatures drop after the first rise. For example, natural climate cycles in the recent geological past indicate that global warming levels of 1– 1.5°C are associated with 10–15% loss of the world's continental ice volume, resulting in sea-level rise of 6–9 meters above the present level. Such levels of warming will be reached in all realistic scenarios within about 10–30 years (Figure 8.3). It will likely take between 300 and 900 years for sea level to rise to those levels, depending on the climate mitigation pathway followed. The shorter timescales will apply if we follow a scenario of business-as-usual emissions, or one with undue further delays in climate mitigation.

Ice-sheet reduction will increase global-mean annual-mean radiative forcing because of reduced reflection of incoming short-wave radiation (sunlight), but the impact will be concentrated in high-latitude regions. Associated seasonal snow and ice reduction, permafrost melt-back, sea-ice reduction, and other fast feedbacks will roughly double the radiative change caused by ice-sheet reduction alone. In such a system of interlinked feedbacks, it doesn't matter where exactly a perturbation starts; eventually all processes develop together. By the time the slow (high inertia) climate feedback of ice-volume loss begins to participate substantially, the ice-sheets are already considerably out of equilibrium with the general climate state (Chapter 6). And this is where we find ourselves today. Satellite and field monitoring of the Antarctic and Greenland ice sheets indicate clearly that ice loss is ramping up in response to the disequilibrium that has grown over the past 200 or so years of anthropogenic climate forcing. Because the ice responds to a large and existing disequilibrium, the only hope we have with aggressive climate mitigation is to slow down the rate of loss; it is unlikely to be stopped on timescales of less than a few centuries, let alone reversed. In consequence, the observed trend of increasing ice loss indicates that we have by now embarked upon an ice-mass loss (and sea-level rise) trajectory that will continue for many centuries, even under aggressive climate mitigation. This cannot be discarded as some concern for the future; a recent detailed evaluation confirmed that the Greenland Ice Sheet already switched into such a state of persistent mass loss 15–20 years ago (King et al., 2020).

Second, let's look at tipping points. While the action of slow feedbacks is a matter of serious concern, it might—at least theoretically—still be weakened to some extent by immediate aggressive mitigation strategies. In contrast, we may expect virtually irreversible climate responses in association with the crossing of climate tipping points, when climate transitions into another stable state (section 2.3). There is growing concern among climate scientists that we are moving perilously close to several major climate tipping points, such as permafrost collapse and associated methane release, rainforest deforestation, and collapse of major portions of the world's great ice sheets.

Because crossing a tipping point causes long-term irreversible change, not even the most aggressive climate mitigation would succeed in bringing climate back to anything like present-day conditions. Instead, variations within a new stable state would be the norm, which will likely mean variations within a climate state that overall has considerably less ice volume, severely reduced permafrost, substantially enhanced methane and $CO_2$ levels because of emissions from permafrost reduction, and a multitude of knock-on effects. Multiple stable states and tipping points are not science fiction, they are a well-known characteristic of natural and social systems (e.g., Beisner, 2012; Milkoreit et al., 2018; Moore, 2018). It is, therefore, a matter of utmost importance that such transitions are avoided. In other words, absolutely no time may be lost with implementing emissions reductions, new emissions avoidance measures, and NETs in accordance with their technological readiness. It is critical that great priority is given to development of new and concept-stage NETs toward the at-scale implementation stage, and that this is paralleled by technology-specific framework development for responsible application and governance.

Finally, the issues of slow responses/feedbacks and tipping points highlight that, in addition to NETs, also SRMs need to be subjected to urgent, full life-cycle assessment, given that we may well need them to slow down warming, at least temporarily, until the aforementioned greenhouse gas mitigation measures have been brought up to speed. We have seen that SRM implementation is associated with profound equality and social acceptance issues. Therefore, international frameworks for agreement and governance need to be developed in parallel with the full life-cycle assessments, to ensure that the world is ready to implement SRMs responsibly once their potentials and downsides have been brought out and weighed up.

# Note

1. As a technical aside: although further investigations may pare down the estimated capacities for OF and NEH2, it is also likely that alternative (yet to be discovered) large-capacity marine NETs might be added. After all, the theoretical capacity of the oceans to capture and store carbon is immense. Thus, the OF and NEH2 contributions in Figure 8.1 might usefully be viewed as generic placeholders for any combination of large-capacity marine NETs, including any that are yet to be discovered.

# Key Sources and Further Reading

Beisner, B.E. Alternative stable states. *Nature Education Knowledge*, 3, 33, 2012. https://www.nature.com/scitable/knowledge/library/alternative-stable-states-78274277/

Bloomberg New Energy Finance, 2019. As in Chapter 3.

Fuhrman, J., et al. From zero to hero? Why integrated assessment modeling of negative emissions technologies is hard and how we can do better. *Frontiers in Climate*, 1, December 4, 2019. https://www.frontiersin.org/articles/10.3389/fclim.2019.00011/full

Fuss et al. 2018. As in Chapter 4.

Grübler, A., et al. Dynamics of energy technologies and global change. *Energy Policy*, 27, 247–280, 1999. https://www.sciencedirect.com/science/article/abs/pii/S0301421598000676

Henderson, G., et al. 2018. As in Chapter 4.

IPCC. 2019. As in Chapter 6.

King, D.K., et al. Dynamic ice loss from the Greenland Ice Sheet driven by sustained glacier retreat. *Communications Earth and Environment*, 1, 1–7, 2020. https://www.nature.com/articles/s43247-020-0001-2.pdf

Lefebvre, D., et al. Assessing the potential of soil carbonation and enhanced weathering through Life Cycle Assessment: a case study for Sao Paulo State, Brazil. *Journal of Cleaner Production*, 233, 468–481, 2019. https://www.sciencedirect.com/science/article/pii/S0959652619320578?via%3Dihub

Lenton, T., et al. 2008. As in Chapter 2.

Lenton, T., et al. 2019. As in Chapter 2.

Liu, Z., et al. Near-real-time monitoring of global CO2 emissions reveals the effects of the COVID-19 pandemic. *Nature Communications*, 11, 5172, 2020. https://www.nature.com/articles/s41467-020-18922-7

Milkoreit, M., et al. Defining tipping points for social-ecological systems scholarship—an interdisciplinary literature review. *Environmental Research Letters*, *13*, 033005, 2018. https://iopscience.iop.org/article/10.1088/1748-9326/aaaa75/pdf

Moore, J.C. Predicting tipping points in complex environmental systems. *Proceedings of the National Academy of Sciences*, *115*, 635–636, 2018. https://www.pnas.org/content/115/4/635

National Academies of Sciences, Engineering, and Medicine. 2019. As in Chapter 4.

PALAEOSENS Project Members. Making sense of palaeoclimate sensitivity. *Nature*, *491*, 683–691, 2012. https://www.nature.com/articles/nature11574

Rau, G.H., et al. 2018. As in Chapter 4.

Rohling, E.J., et al. 2013. As in Chapter 6.

Rohling, E.J., et al. 2018. As in Chapter 2.

Rohling, E.J., et al. 2019. As in Chapter 6.

Schmidt, H.P., et al. 2019. As in Chapter 4.

Shepherd, J.G., et al. 2009. As in Chapter 4.

# 9
# Conclusions

To wrap up, let's step back from details and focus entirely on the most important high-level messages of the curves in Figures 2.5 and 8.3. These clearly illustrate the following:

1) **A combined portfolio of emissions reduction/avoidance and NETs can still lead to conditions that conform to the 1.5°C limit of the Paris Agreement.**

2) **Neither emissions reduction and avoidance nor NETs can do it in isolation.** Emissions reduction alone would have to reach draconian levels of 7–8% reduction per year, which seems an economic impossibility (Figure 2.5). NETs alone cannot do it either, as shown by the 0% lines in Figure 8.3. Using NETs for offsetting emissions, rather than eliminating those emissions, is a particularly bad practice because neither method then makes an impact where it should.

3) **A promising way forward combines 2–4% per year emissions reduction with an ambitious and diverse portfolio of NETs.** Implementation of this NETs portfolio needs to start immediately with methods that are at high technological readiness, along with parallel development and implementation of methods of lower technological readiness. This recommendation is fundamentally similar to those from more complex assessments, which also indicate that limiting warming to 1.5°C requires far-reaching emissions reduction pathways combined with NETs deployment to draw down 3–5 GtC per year (Fuhrman et al., 2019).

Further conclusions from the matter reviewed in this book are as follows:

4) **Solar radiation management (SRM) may provide important services for bringing under control the initial warmings until about 2050** (Figure 8.3). This short timeframe implies a firm focus on methods

that are ready to go immediately. For example, space mirror initiatives lose a lot of relevance in this context because they will require too much time to get going and will accordingly miss the time window in which they are most needed. Stratospheric aerosol injections, in particular, seem to be approaches of choice, given their high potential impact and relatively high technological readiness (Figure 5.6). Marine cloud brightening seems promising too, but would need to overcome a tendency for regional limitation and potential impacts of weather systems.

5) **Adaptation must be part of the response to climate change.** A range of changes related to ocean warming and land-ice reduction—and resultant sea-level rise—has become inevitable already, bearing major, ever-increasing risks to food resources and space for living and livelihoods. These inevitable changes require adaptation, which must not be confused with "elective" adaptation to changes in rapid processes that can and should be avoided by appropriate emissions reduction and NETs and/or SRM application.

6) **To ensure economic feasibility and sustainability, it is critical that efforts** in emissions reduction/avoidance, NETs, SRM, and adaptation **are focused where they are truly needed, and not wasted on fixing issues that could and should be avoided.** For example, using NETs to offset emissions that could be avoided is a waste of effort and resources. Similarly, adaptation to processes that can still be stopped is a waste of effort and resources. It is essential that the right solution is mapped onto each problem.

7) **New governance, regulation, and legal frameworks must be developed in real time, in parallel (and flexibly adapting) to technological developments and upscaling of the various techniques.** Continued public and focus-group pressure will be needed to keep regulating bodies on track and sufficiently ambitious. Legal challenges to laggards, offenders, and obstructions will help create and maintain momentum. Civil liability cases against major emitters can drive self-regulation, and even failed legal challenges will help to articulate climate change as a legal and financial risk.

\* \* \*

The 2019–2020 novel coronavirus crisis has already proven that, given the right incentives, the massive industrial machine can re-task extremely quickly under the right circumstances. Automobile manufacturers switched to producing ventilators, smaller manufacturers switched their efforts to personal protection equipment, and so on. This is a promising sign for the industrial re-tasking that will be involved in rebalancing climate. By mid 2020, the coronavirus crisis has also demonstrated the economic and personal hardships associated with shock implementation of a 5–10% emissions reduction. Even if we maintained a 10% emissions reduction in this way over a few years, which would seriously risk a global economic collapse, we would not be reducing the amount of carbon in the atmosphere. In fact, some 9 billion tons of carbon emissions per year would continue. In consequence, cumulative carbon amounts in the atmosphere would still increase rapidly and, therefore, so would global temperature. It really shines a bright spotlight on the scale of humanity's gigatons challenge.

This book has gone through a broad portfolio of measures by which we may still meet the gigatons challenge. But the window of time left to allow us to stay within the Paris Agreement's 1.5–2 °C warming limit is closing rapidly. We need prioritization of this challenge at all levels: emissions reduction; new emissions avoidance; NETs; and very likely also SRM. And we also need to prioritize development of appropriate legal, social, and governance frameworks to ensure rapid but responsible implementations, as well as monitoring and regulation.

Implementation of such a broad and diverse portfolio of measures requires visionary leadership, born from a capacity to recognize that this change will future-proof society. After all, it will bring new industrial sectors; technology development; new employment solutions; potential for integrated solutions across energy self-sufficiency, carbon removal, and food security; financial, public, and environmental health benefits; and a wide variety of opportunities in both internal and export markets.

And, to circle back to lessons from the coronavirus crisis: it has long been known that humanity's current pathway of unbridled expansion, unchecked mobility, and living at the expense of nature—rather than sustainably with nature—is an important driver behind the emergence and devastating spread of zoonotic diseases. A global societal-economic shift to a sustainable model, with serious efforts at rebalancing climate as one of its key attributes, will help with reducing the emergence and spread of such diseases, and thus reducing their threats to public health and the global economy.

**The time to act is now.**

# Climate Feedbacks

This appendix provides a summary look at feedback processes in the climate system and highlights which are rapid and which are slow.

Feedbacks are processes that are activated in response to an initial disturbance. They can be positive, which means that the process amplifies the initial disturbance. Or they can be negative, in which case the process dampens the initial disturbance. We all know feedback processes from experience. A squealing microphone is an example of positive feedback. A car's shock absorber is an example of negative feedback.

The water vapor feedback that we encountered before is a powerful positive feedback in the climate system, and it operates nearly instantaneously. There are many more of these so-called fast feedbacks, which operate over timescales of less than a year to multiple decades. Some, like water vapor and also cloud-formation feedbacks, are called rapid adjustments rather than feedbacks by WG1 of the IPCC AR5 (2013), to distinguish these from other effects that modify the planet's balance of radiative forcing following changes in surface temperature. These are the "real" fast feedbacks, and they include seasonal snow and ice variations that affect regional reflectivity to ISWR; influences of aerosols of mineral dust blown from deserts and affecting cloud formation; and variations in natural methane release from swamps, bogs, permafrost, and potentially the faster portions of methane hydrate decomposition (Figure A1.1).

Next, there are medium to long-term feedbacks, which operate on time scales from many decades to several centuries (Figure A1.1). These include vegetation-zone migrations that affect land-surface reflectivity and can also modify dust availability (for example, when a desert becomes covered in vegetation, or vice versa); large-scale methane hydrate decomposition; the faster components of land-ice change, in particular where these ice sheets enter the ocean; and faster components of change in the operation of the global carbon cycle, which affect greenhouse-gas uptake or release.

Finally, there are very slow processes that operate over thousands to millions of years (Figure A1.1). These include the slower components of land-ice change and the global carbon cycle. Notably slow components of the carbon cycle that operate over tens of thousands to millions of years are those related to weathering, plate tectonic adjustments including mountain formation and volcanicity, and biological evolution of completely new vegetation types.

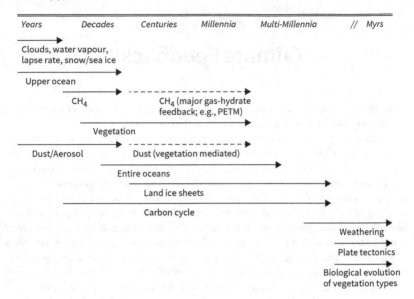

**Figure A1.1**  Overview of climate feedbacks and their timescales.

*Source*: From Rohling, E.J. *The climate question: natural cycles, human impact, future outlook*. 162 pp., Oxford University Press, New York, 2019.

## Key Sources and Further Reading

Intergovernmental Panel on Climate Change (IPCC), 2013. As in Chapter 2.

Rohling, 2019. As in Chapter 2.

Rohling, et al., 2012. As in Chapter 2.

Rohling, et al., 2018. As in Chapter 2.

# Indicative Future Projections for Carbon Removal by NETs

Based on Fuhrman et al. (2019), each NET's development is represented by a sigmoidal, so-called logistic-style, function in which initial exponential growth changes to similar slowdown of growth and then settles at a constant level of carbon removal. The function shape used is

$$f_{(t)} = \frac{\alpha}{1 + e^{-C\left(\frac{t}{\tau} - \kappa\right)}}$$

Here, $\alpha$ is the complete amplitude achieved once enough time has passed for the function to stabilize; $t$ is years into the future from 2020, $\tau$ is the timescale from start to stabilization of the function at amplitude $\alpha$; $\kappa$ is a simple translation constant equal to 0.5 for all functions, which is set so that growth commences from the first year calculated (2020); and $C$ is a simple constant equal to 10 for all functions, which is set so that amplitude $\alpha$ is achieved at time $\tau$. The calculated value $f_{(t)}$ is the growth through time of the carbon removal value for each NET considered (in GtC/y).

In this exercise, timescale $\tau$ for each NET is determined in relation to its TRL value, as per Table A2.1, except in the cases of DACCS and EW (Table 8.1). For DACCS, $\tau$ is

Table A2.1. TRL dependence of the mean $\tau$ values used for NETs projection.

| TRL | Mean $\tau$ |
| --- | --- |
| 9 | 30 |
| 8 | 40 |
| 7 | 50 |
| 6 | 60 |
| 5 | 70 |
| 4 | 80 |
| 3 | 90 |
| 2 | 100 |
| 1 | 110 |

set to 100 years, roughly based on growth curves from integrated assessment models (Fuhrman et al., 2019). For EW, $\tau$ is set to 150 years, roughly based on a 100 times acceleration of natural weathering rates of 0.021% per year (Lefebvre et al., 2019). Uncertainties in amplitude $\alpha$ and timescale $\tau$ are approximated using simple uniform distribution ranges (Table 8.1). For $\alpha$, a mean GtC/y value and surrounding ±GtC/y values are determined from the synthesis ranges presented in the top panel of Figure 4.12. Uncertainty ranges for $\tau$ are simply set to ±50% of the mean value.

## Key Sources and Further Reading

Fuhrman et al., 2019. As in Chapter 8.
Lefebvre et al., 2019. As in Chapter 8.

# Index

For the benefit of digital users, indexed terms that span two pages (e.g., 52–53) may, on occasion, appear on only one of those pages.

Tables and figures are indicated by *t* and *f* following the page number